# PERSPECTIVE

## FOR ART STUDENTS

BY

### RICHARD G. HATTON

AUTHOR OF

"ELEMENTARY DESIGN," "FIGURE DRAWING AND COMPOSITION"

**With 208 Diagrams**

British Library Cataloguing-in-Publication Data
A catalogue record for this book is available from the
British Library

# Technical Drawing and Drafting

Technical drawing, also known as 'drafting' or 'draughting', is the act and discipline of composing plans that visually communicate how something functions or is to be constructed.

It is essential for communicating ideas in industry, architecture and engineering. The need for precise communication in the preparation of a functional document distinguishes technical drawing from the expressive drawing of the visual arts. Whereas artistic drawings are subjectively interpreted, with multiply determined meanings, technical drawings generally have only one intended meaning. To make the drawings easier to understand, practitioners use familiar symbols, perspectives, units of measurement, notation systems, visual styles, and page layout. Together, such conventions constitute a visual language, and help to ensure that the drawing is unambiguous and relatively easy to understand.

There are many methods of constructing a technical drawing, and most simple among them is a sketch. A sketch is a quickly executed, freehand drawing that is not intended as a finished work. In general, sketching is a quick way to record an idea for later use, and architects sketches in particular (in a very similar manner to fine artists) serve as a way to try out different ideas and establish a composition before undertaking more finished work. Architects drawings can also be used to convince clients of the merits of a design, to enable a building constructer to use them, and as a record of completed work. In a similar manner to engineering (and all other technical drawings), there is a set of conventions (i.e particular views, measurements, scales, and cross-referencing) that are utilised.

As opposed to free-sketching, technical drawings usually utilise various manuals and instruments. The basic drafting procedure is to place a piece of paper (or other material) on a smooth surface with right-angle corners and straight sides – typically a drawing board. A sliding straightedge known as a 'T-square' is then placed on one of the sides, allowing it to be slid across the side of the table, and over the surface of the paper. Parallel lines can be drawn simply by moving the T-square and running a pencil along the edge, as well as holding devices such as set squares or triangles. Other tools can be used to draw curves and circles, and primary among these are the compasses, used for drawing simple arcs and circles. Drafting templates are also utilised in cases where the drafter has to create recurring objects in a drawing – a massive time-saving development.

This basic drafting system requires an accurate table and constant attention to the positioning of the tools. A common error is to allow the triangles to push the top of the T-square down slightly, thereby throwing off all the angles. Even tasks as simple as drawing two angled lines meeting at a point require a number of moves of the T-square and triangles, and in general drafting this can be a time consuming process. In addition to the mastery of the mechanics of drawing lines, arcs, circles (and text) onto a piece of paper – the drafting effort requires a thorough understanding of geometry, trigonometry and spatial comprehension. In all cases, it demands precision and accuracy, and attention to detail.

Conventionally, drawings were made in ink on paper or a similar material, and any copies required had to be laboriously made by hand. The twentieth century saw a shift to drawing on tracing paper, so that mechanical copies could be run off efficiently. This was a substantial development in the drafting process – only eclipsed in the twenty-first century with 'computer-aided-drawing' systems (CAD). Although classical draftsmen and women are still in high demand, the mechanics of the drafting task have largely been automated and accelerated through the use of such systems. The development of the computer had a major impact on the methods used to design and create technical drawings, making manual drawing almost obsolete, and opening up new possibilities of form using organic shapes and complex geometry.

Today, there are two types of computer-aided design systems used for the production of technical drawings; two dimensions ('2D') and three dimensions ('3D'). 2D CAD systems such as AutoCAD or MicroStation have largely replaced the paper drawing discipline. Lines, circles, arcs and curves are all created within the software. It is down to the technical drawing skill of the user to produce the drawing – though this method does allow for the making of numerous revisions, and modifications of original designs. 3D CAD systems such as Autodesk Inventor or SolidWorks first produce the geometry of the part, and the technical drawing comes from user defined views of the part. This means there is little scope for error once the parameters have been set.

Buildings, Aircraft, ships and cars are now all modelled, assembled and checked in 3D before technical drawings are released for manufacture.

Technical drawing is a skill that is essential for so many industries and endeavours, allowing complex ideas and designs to become reality. It is hoped that the current reader enjoys this book on the subject.

# PREFACE

"To pore over all these matters Paolo would remain alone, seeing scarcely any one, and remaining almost like a hermit for weeks and months in his house without suffering himself to be approached." So wrote Vasari of Paolo Uccello.

To many since Uccello's time this "most elegant and agreeable art," as the author of the "Jesuit's Perspective" regarded it, has had fascinations, and much midnight oil has been burned by its votaries.

The "Jesuit's Perspective," a seventeenth-century work, was reprinted for the seventh time late in the eighteenth century, and its errors and shortcomings seem to have stimulated the sounder mathematicians of the time to literary activity.

These writers carried the art to its perfection, so far as theory was concerned. They found vanishing-points for lines in all possible positions, no matter how curiously inclined to the picture.

To the author of the "Jesuit" this advanced part of the subject never once occurred; he even marks as accidental, vanishing-points on the horizon other than the point of sight. With only parallel perspective, it is small wonder that so tireless an author should seek to increase the usefulness of his science by facilitating its manipulation, since he could not extend its bounds.

Perhaps the most remarkable instance is that which he borrows from "the sieur G. D. L." His object is always to get the result quickly, without confusion, without long working lines, and with as few of them as possible. Hence, in this case, he avoids the use of a line elevation for his heights, and avoids also coming continually down to his ground-line for his dimensions He marks six feet on his ground line as a scale, and divides the first into inches. This scale he runs back to the point of sight. Provided thus with a regularly diminishing scale, if at any point on the ground he wishes to take or raise a measurement, he has his scale diminished to its due extent at the very point. He merely draws a level line at the point, and this, as it crosses the scale, indicates where he has to set his dividers. Then, to avoid having a distance-point off his paper, or at least a great way off, he measures his distances within by a scale only one-fourth of the scale for the widths and heights of which we have just been speaking. His distance-point is quite near. It is, in fact, a fractional distance-point measuring fourfold; and how accessible it is the student will very soon find. Furthermore, he uses the distance-point as it is used in

Fig. 52, page 76, which, in fact, is also to be placed to his credit.

Nowadays, the architects are the only people who can make perspective drawings; the art-student usually only draws impossible geometrical figures by a roundabout process. With the great change for the better in the syllabus of the Examinations of the Board of Education there is a probability of more useful study being followed. The syllabus does not, however, touch the practical side, such as designers would employ, and the student would be advised not to omit that kind of study.

The secret of success in such a subject as perspective lies in the student having the whole range planned out, so that he knows where his study actually comes to an end. Without such tabulation he must always have the feeling that with every new problem some unheard of difficulty will present itself. Under such a strain he never gets fairly face to face with what he has to do. This tabulation must be done by the student himself; he must plan the whole ground out, taking, perhaps, the rules given on pages 267 to 270 as a rough guide. He must settle how many figures he regards as sufficient to cover all difficulties—a square, a triangle, a circle, a curve; or, again, a cube, a pyramid, an object with legs. Then there are the difficulties of position—parallel, at an angle, near, remote, high up, low down. Then the difficulties of tilted position.

All these matters are capable of limitation, beyond which it is waste of time to travel.

The following pages have been put together with a view to their being *read*, so that the student is advised to gain some general acquaintance with the subject before *working* out problems. Before he begins working, he should certainly sketch solutions, doing the work by free-hand instead of by rule and compass.

# CONTENTS

# Contents

# THE NEW SYLLABUS OF THE BOARD OF EDUCATION

IS AS FOLLOWS:

The Candidate will be expected to show—

(*a*) Skill in using instruments and working ont one or two problems accurately, and

(*b*) Evidence of ability in the ready application of the rules of Perspective to the representation of objects views of buildings, landscapes, etc., by freehand sketches in pencil, ink, or sepia.

There will be 1st and 2nd class passes. A 2nd class success will be accepted for the Elementary Drawing Certificate, and a 1st class for the Art Class Teacher Certificate.

Candidates must qualify in (A) and (B).

(A) Representing in perspective from plan and elevation or from specification, simple solids or objects of plane or curved surfaces having one line or surface on, or parallel to, the ground-plane.

Drawing and measuring lines inclined to the horizontal, and contained in vertical planes inclined to the picture-plane.

Drawing figures or solids in perspective, some of whose leading construction lines are horizontal and the others contained in vertical planes at right angles to the horizontal lines, *e.g.* a cube with one edge horizontal, and one face making a given angle with the ground.

Drawing solids having plane or curved surfaces in oblique positions, and all their constructive lines inclined to the ground, such representation being limited to problems which can be solved by the treatment of an oblique plane and perpendiculars thereto.

Drawing reflections of solids in plane mirrors, horizontal or vertical.

Drawing shadows of lines, surfaces or solids, rectilineal or curved, upon any specified planes and on surfaces of single curvature, by natural or artificial light.

(B) Finding and describing from views given in perspective the actual dimensions, positions, and other particulars respecting the objects represented under the conditions of any of the foregoing classes of subject (or in the case of shadows and reflections, ascertaining the position of the source of light, reflecting surface, etc.).

Idicating how change of position, say of the spectator, of the object, or of the source of light, etc., affects the representation of the given objects, etc.

Indicating effects of distance on the appearance of objects, shadows, etc.

Pointing out and correcting errors in perspective in respect of given subjects.

# ADDITIONAL NOTES TO THE EDITION OF 1910.

## I. Practical Application of Perspective.

Erect a sheet of glass, say about 30 × 20 inches, vertically upon a table. Place on the table a strip of paper 1 foot wide and, say, 6 feet long, marked across into squares. Let the narrow end of the paper touch the glass ; 12 inches in front

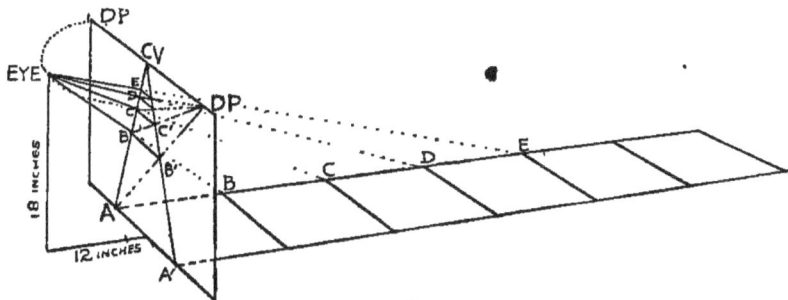

Fig. i.

of the glass make a station point on the table, and erect a rod, say, 18 inches high, the summit of which is to be the eye.

Placing your eye to the top of the rod, and looking at the paper, draw, in water-colour, upon the glass, the image of the squares on the paper.

Fig. i. shows this arrangement. Fig. ii. shows the working upon the picture. What do we learn from these ? That our strip of six squares becomes reduced in vision to a form which has its sides converging to C.V.; that the front edge of the ·first square AA' touches our picture, and is there actual; that the first square, AA'BB', is very distorted in the perspective view; that if we place a Distance Point, D.P., as far along the horizon as the eye is from the C.V., we can measure the limits of the squares backwards. We find, in short, that actual and geometrical perspective correspond.

The square CD is free from distortion, and we come to the conclusion that our picture need not begin lower than,

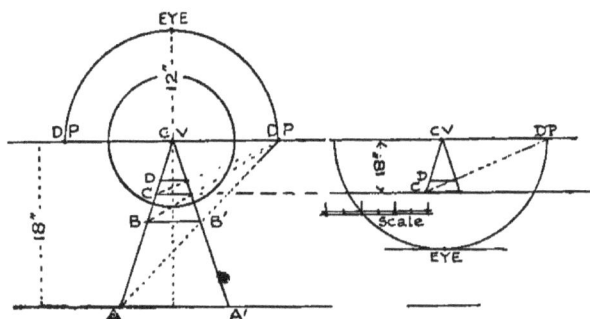

FIG. ii.

say, the line C. If we measure the line CC' on our glass, we find it to be 4 inches; the 12 inches of AA' have been reduced to 4 inches. If, therefore, we start our picture at the level of C, we have to adopt a scale of 4 inches to 1 foot. We find that this new picture line through C is 6 inches below the horizon on our glass, or 1 foot 6 inches according to the scale of 4 inches to 1 foot. Applying the same scale to the distance of the eye before the picture, we may now call it 3 feet, for 12 inches on the scale of 4 inches to 1 foot is 3 feet. This indicates a convention in ordinary geometrical perspective; for the eye really remains only 12 inches before the picture.

We see by our experiment how it comes about that the

measurements upon the picture line are drawn to a scale, and are yet talked of as actual.

Consider now Fig. iii. This is a landscape. The horizon is 6 inches above the lower edge of the picture. How far before us is this front edge, A ? To ascertain that we must decide how far the spectator is to stand before the picture to get properly into its perspective. The picture is 24 inches wide. Suppose we view it at twice that distance away = 4 feet. We must also decide whether the spectator is standing

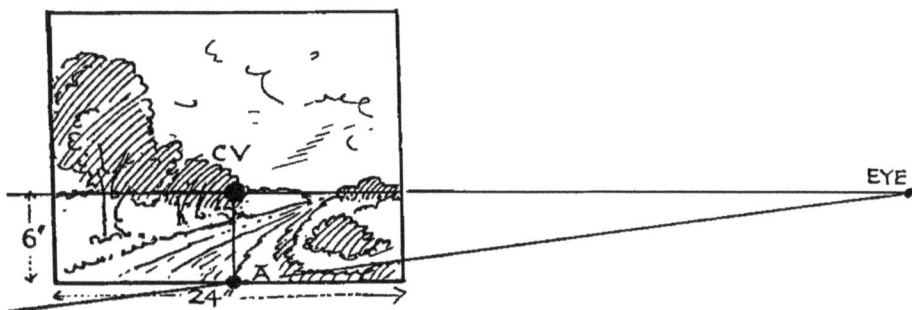

FIG. iii.

on the level of the ground. Assume that he is, and that his eye is 5 feet above the ground. To find the distance that the middle of the lower edge of the picture, A, is from the spectator—how far A is along the road—we must continue the lines B and C till the space BC becomes 5 feet. We can calculate this readily. If 4 feet gives a drop of 6 inches from C.V. to A, it will take 40 feet to give a drop of 5 feet, or 36 feet to give a drop of 4 feet 6 inches, which is the space below the edge of the picture. Therefore, had we drawn the picture line 5 feet below C.V., we should have had to vanish a distance of 36 feet inwards to get to point A.

We see, therefore, that the picture which we paint begins many feet away from us. But if we want the spectator to feel very much in and among the scene we depict, we introduce objects or figures which are standing on nearer ground than that included in the picture (Fig. iv.). Many portraits, and all half or three-quarter figures are of this order.

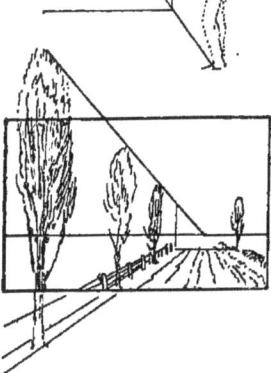

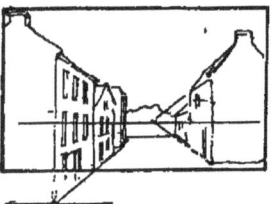

As a general rule, one should stand away from an object fully three times its height, or greatest dimension. From a picture we usually stand not nearer than twice its diagonal from it, perhaps not nearer than three times that measurement. Sometimes artists paint their pictures from points nearer to the picture than the spectators generally stand. When this is the case there is some loss of reality on the spectator's part. If the picture is one of rigid perspective, as in architectural subjects, rooms, streets, and the like, the outer parts of the picture look distorted, till the spectator gets into the correct position. For geometrical perspective is a faultless science, and distorted pictures are only such so long as they are not viewed at the proper distance.

Fig. iv.

It is consequently a great mistake · to draw the perspective from a point impracticably near. Dürer did so in his beautiful print *St. Jerome in his Cell.* The loss of reality is very great in that fine design.

It is better to take a distance too remote than too near. Commonly it is said that comfortable vision is limited to

30° around the central visual ray—60° in all. I find 20°
to 25° more nearly the correct range than 60°.

## II. INVERSE PERSPECTIVE.

Inverse perspective is very important to the artist. He
generally sketches his design purely by freehand, leaving the
laws of perspective alone. By so doing he can arrange his
matter more artistically. But if he wants to test his work,
he must do so inversely.

Many of the inverse questions set at Examinations can be
readily solved by persons who have a grasp of the subject
such as is covered by this book, but some questions depend
upon simple extensions of the subject, which, in former editions,
were not dealt with.

All one's troubles in perspective come from following
stereotyped methods, instead of mastering principles.

In the previous note it has been shown how the front edge
of the picture can be taken back to any convenient position.
In a word, we can have as many ground lines to the ground,
or picture lines to other planes, as we see fit. Refer to Fig. i.
again. Upon the glass picture plane we can choose the line

BB', or CC', or which-
ever we like as our ground
line. In Fig. ii. we chose
CC'. All that happens is
that the scale alters.

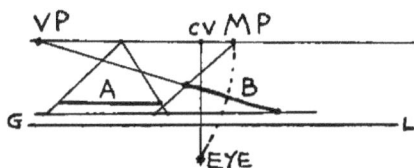

FIG. V.

A simple question set
in 1908 depended upon this : "Two lines of equal length
are given, also Hor., C.V., and G.L. find position of Eye."
The eye will obviously be under C.V. The clue to its position
is given by the slanting line B, whose V.P. can be found.
Now one can only get to the eye by means of V.P.'s and

M.P.'s. The ground line does not help us, because B is away from it, and what troubles us is the distance from B to the G.L. We therefore assume a new or casual G.L. through the first point of B. What we want to get is the M.P. of B, for one V.P. alone will not enable us to get to the eye. We know B = A, so bring forward A to the new G.L. Take the size thus given, and set off against B, and carry the measuring line through the end of B, and so find M.P.

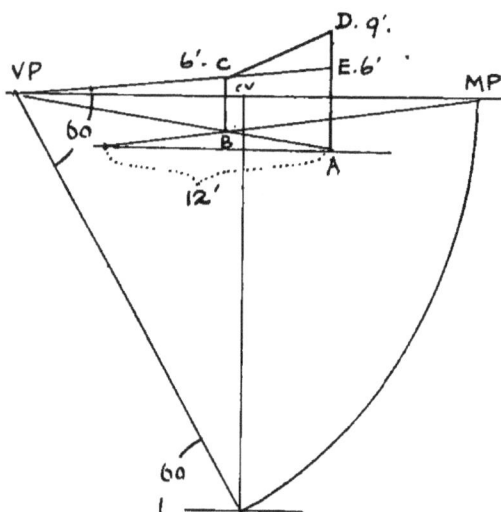

A similar question (April, 1906) is where a four-sided figure ABCD, · is given. We are told that AB = 12 feet, AD = 9 feet, and BC = 6 feet, and that the figure is at 60° to the left. We have to find horizon, C.V., and eye, and to get these we must find V.P. and M.P. of AB. Had the top line, CD, been parallel to AB, we could at once prolong AB and CD till they met. We have to find such a level line. BC is 6 feet, AD is 9 feet. Divided into six or nine parts, they both yield their scales. It will suffice, however, to divide AD into three equal parts. We thus find E 6 feet above A. EC is then parallel to AB, and we can find V.P. Now to find M.P. Draw a new ground line through A ; on it mark 12 feet. that is, double AE just found. Carry the line back through B, and find M.P. AD and BC are given vertical in order

Fig. vi.

that we may know how to draw the horizon, for we have no other indication of the level.

Other problems depend upon what we may call the *casual measuring points*. Lines are measured by obtaining other lines to cross their terminations. All this is dealt with on page 71. We find that one M.P. (really one pair, as shown in Fig. 50) will give us on the picture line the real size which we are getting on the line which is vanishing. This M.P. is the M.P. above all others. But any point will serve to *repeat* distances. Consider Fig. vii. There ABC are two equal

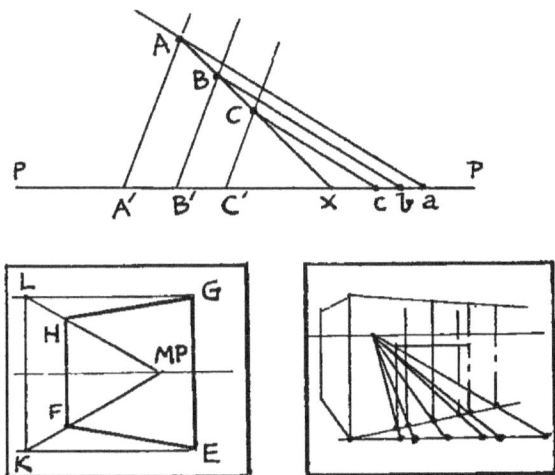

F<small>IG.</small> vii.

dimensions upon the same line. Their proper measuring lines will be AA', BB', CC'. A'B' = AB, for AA'X is an isosceles triangle. But any parallel lines through A, B, and C will yield equal dimensions on PP. Thus we obtain *c*, *b*, and *a*, and though *ab* is less than AB, yet *ab* = *bc* just as AB = BC.

Problems to be worked in confined spaces are usually to be done by this casual M.P. Thus EFGH (Fig. vii.) is a rectangle :

find the horizon, no working lines to be beyond the limits given. If we could continue EF and GH till they met, we should at once get V.P. and so the horizon through it. That not being allowed, we must find either another V.P. or an M.P. We cannot find the true M.P. but we can find a casual one. By having EF and GH equal, we are enabled to do this. We create two picture lines, one through E, the other through G. We mark *any* distance, EK, and make GL equal. Draw KF and LH. These are over one another, and will meet at a casual M.P. on the horizon.

Questions are set in which half of an object, a house, a desk, is given, and the other half has to be got. Here, again, the real M.P. not being obtainable or allowed, we use a casual M.P. An example of such a problem is also given in Fig. vi.

If *the* measuring point is wanted, some clue must be given. This generally is that a receding line is given as equal to one parallel to the picture. We had an example in Fig. v. But sometimes the line which gives the clue is vertical. A question (June, 1907) was similar to that of the rectangle EFGH in Fig. vii. But it asked for *the* measuring point. The clue was given in there being a border to the rectangle. The border was equal all round, and so an actual measurement could be made.

Among other means of working inverse questions, the use of the V.P. of diagonals must not be overlooked. A question of a table (June, 1907) depended upon the use of this V.P. Students who are afraid of the " other " planes—oblique, vertical ascending, etc.—lose valuable aids on occasion. Vertical planes should most certainly be thoroughly mastered. In all cases *remember where the perpendiculars to the planes go.* Do they vanish? and if so, where?

# SET OF EXERCISES

## FORMING A COURSE OF 18 LESSONS.

*The references are to pages, or figures, in the book, where help may be gained in dealing with the problems. P. 66 means page 66, and f. 66 means figure 66. Where a figure is referred to, it is intended that the whole comment upon that figure should be considered.*

LESSON 1. (i) Erect a sheet of glass in the manner described on page xv. above, and upon it draw a cube. The cube has 12 inch sides and stands on a horizontal plane. It is 2 feet from the glass, and one side of it is parallel to the glass. On the other side of the glass arrange the position of the eye, 18 inches above the horizontal plane on which the cube stands, and 12 inches before the glass, and opposite the middle of the cube. Place your eye at the position just found, and view the cube through the glass. Keeping the eye steady drawn on the glass the cube as seen. Mark on the glass the horizon and the C.V., noting that the receding lines of the cube are directed towards C.V., which is thus their V.P. ,

(ii) Make a geometrical drawing of the same subject, using the measurements full-size. [This drawing should give exactly the same result as the preceding experiment.] (References f. 35, f. 43, f. 53, f. 91, f. 97).

(iii) Distance (of eye before P.P.) 3 feet. Height (of eye above horizontal plane) 18 inches. A cube 1 foot sides lies on H.P. immediately before the spectator with one side parallel to P.P. *and touching it. Scale* 4 inches = 1 foot. [This drawing should give exactly the same result as the preceding exercise. The two exercises differ only in the manner of the statement.]

LESSON 2. (i) Using the glass as in Lesson 1 (i), draw the present book lying on a table before you. The book is 8 × 5 × 1¼ inch thick. Place it 2 feet 6 inches from you. Place the glass P.P. 1 foot from you, and 1 foot 6 inches before the book. Let the book lie angularly, so that the nearest corner is immediately in front (that is neither to right nor left of the middle line) and the long and short edges are both at 45° to the glass P.P. Height of eye above the table, 14 inches. Add the curves and details by freehand.

(ii) Scale ⅜ inch = 1 inch. Height 14 inches. Distance 32 inches. A book 8 × 5 × 1¼ inch thick lies on H.P., its nearest corner touching the P.P. immediately before the spectator. The edges of the book recede at 45° to right and left. Complete by adding curves, etc., by freehand. [This exercise will give the same result as the preceding experiment]. (References, f. 25, f. 35, f. 43, f. 45, f. 48, f. 50, f. 58, f. 107).

LESSON 3. (i) Height, 5 feet. Distance, 12 feet. Scale, ½ inch = 1 foot. Four strips of cloth lie on the ground. Each is rectangular, 6 × 2 feet. They lie parallel to one another with spaces of 18 inches between them, but so placed that the second, third, and fourth successively have their nearer ends level with the middles of the first, second, and third. The long edges are to be perpendicular to the P.P., and that strip which is farthest on the left touches the P.P. with its nearest corner 4 feet on left of centre.

(ii) The same as the preceding exercise, but having the long edges of the strips at 60° towards the left, and the piece farthest on the left touching the P.P. by one corner, 4 feet on the left. (References, f. 53, f. 54).

LESSON 4. (i) Height, 5 feet. Distance, 12 feet. Find the locations of the following points. A is 5 feet on left, 2 feet within the picture. B is 2 feet left, 6 feet within. C is 3 feet right, 9 feet within. D is 6 feet right, 20 feet within. Upon each of these points raise perpendiculars 8 feet high, giving A', B', C' and D'. Join AB, BC, CD and A'B', B'C', and C'D'. Find the V.P.'s and M.P.'s of AB, BC, and CD, and measure them. Scale ½ inch = 1 foot. (References, f. 49, f. 52).

(ii) Distance, 12 feet. Scale, ½ inch = 1 foot. Point E is 4 feet above the level of the eye; it is 3 feet on right and 5 feet beyond the P.P. E is the upper end of a vertical line, EF 7 feet long. EF is axis of a pyramid whose base is a square 5 feet sides. One diagonal of the square base is parallel to P.P. (References, f. 79, f. 140).

LESSON 5. (i) Subject given on p. 82. Height, 5 feet. Distance, 10 feet. Scale ½ inch = 1 foot. The lowest step is 10 feet square. The subject being completed, consider the upper surface of the uppermost step to be the base of a pyramid 8 feet high. Draw the pyramid. This exercise is in setting back forms one beyond another. Many errors in the working of perspectives are due to confusion in this matter, which is really simple.

(ii) Subject given on p. 87. Add the pyramid as in the preceding exercise. It would be a useful variation for the student to use a separate picture line for each step, and for the pyramid.

LESSON 6. (i) Height, 5 feet. Distance, 12 feet. Scale, ½ inch = 1 foot. Point A on ground 6 feet on left, and in P.P. is beginning of a line at 35° with P.P. towards right. This line is near side of a road 20 feet wide. At a point 30 feet along the road from A, the road changes its direction, and then runs at 60° with P.P. towards right. After continuing another 30 feet, it again changes its direction and runs directly perpendicular to P.P. (Work the near side of the road first. The problem requires a space extending from 11 inches on left of C.V. to 18 inches on right of it). Add cart ruts, etc. (Reference, f. 49).

(ii) Add to the above a number of trees. Place anywhere 6 or 8 points. These are bases of trees all 30 feet high. (References, f. 44, f. 45).

[Both these exercises deal with long measurements. The student will not attempt to use such a height as 30 feet, but will probably use 10 feet, and, having erected 10 feet at each tree, will merely triplicate the size obtained by using the dividers. In the case of the road, where the span from V.P. to eye is sometimes very great—beyond the span of the compasses—he will draw a line parallel to the horizon, say 2 inches above the eye, and will there strike his arcs and get points like M.P.'s, carrying lines through these points, from the eye up to the horizon. This procedure we follow also when a V.P. is inaccessible. We then know the direction of its *vanishing parallel*, and so can get M.P. We, of course, merely use the geometrical method for dividing one line in the same proportion as another.]

LESSON 7. (i) Scale, full size. Distance, 18 inches. A large cotton-reel stands on one end, its centre 3 inches on left, and in the plane of the picture. The drum of the reel is 3 inches long, $1\frac{1}{2}$ inch in diameter. The flange, at either end, may be represented by a truncated cone, expanding from $1\frac{1}{2}$ inch diameter to 2 inches diameter, the axis being half an inch. The reel has thus a total length of 4 inches. Draw the reel in perspective, standing on a plane 7 inches below the eye.

(ii) Draw the same reel when its axis is parallel both to the horizontal plane and the P.P. and 3 inches below the level of the eye, the nearer end of the axis being 3 inches on right, the axis entirely in the plane of the picture. [*Note.*—This view will appear distorted except when viewed exactly from the correct position. The difficulty of viewing the picture from the right position leads draughtsmen to avoid such attitudes, in preference for angular positions which more readily look correct.] (References, f. 50, f. 162.)

LESSON 8. (i) A millstone is 6 feet in diameter, 1 foot thick, and has a square hole in its centre, the diagonals of the square being 1 foot. Draw in perspective when the millstone lies flat on the ground, the nearest point of its circumference to the P.P. being 4 feet left, 2 feet within. One diagonal of the square hole is parallel to the P.P. Scale $\frac{1}{2}$ inch = 1 foot. Height 6 feet. Distance 12 feet.

(ii) The same millstone stands on its edge on the ground. Its nearer circular face is in a vertical plane receding towards the left at 60° with P.P. The nearest point of the circumference to the P.P. is 6 feet right, 2 feet within. One diagonal of the square hole recedes upwards at an angle of 40° to the G.P. (References, f. 60, f. 149.)

LESSON 9. (i) Hitherto the exercises have been of subjects upon the ground-plane, or upon some other horizontal plane. Our working has been chiefly upon the horizontal planes, now and then, when getting a height we have used vertical planes, almost without knowing that we did so. But it is unwise for the student longer to neglect the study of the planes which are not horizontal. The use of V.P's "up in the air" and "down below the ground" is very serviceable, and saves a great deal of trouble. Moreover, the true foreshortening is so much more thoroughly exhibited. These V.P's are often called *accidental* V.P.'s (A.V.P.)—a term used in this book. They are really no more accidental than those on the horizon.

In this lesson the student should learn that there are only seven possible positions of planes (p. 27), of which one, that which is parallel to P.P., has no V.P.'s, and does not recede. There are thus six which have vanishing lines or "horizons," V.P.'s, M.P.'s, and so on.

The lesson should be occupied in studying the six planes, or six positions of planes, using the sheet of glass, and noting the positions of the vanishing lines. Pages 29 to 48 deal with these matters.

Let the following be the Exercises. On six different papers prepare the following : Mark four points in any positions. Mark one C.V., and the others A, B, and C. Let the points not be over one another, nor directly right and left of one another. The positions need not be the same on all the problems.

(i) Draw the V.L. of *horizontal* planes (the horizon). Join AB and AC and BC. These are three lines in a horizontal plane, and together form a triangle in the plane. Find the V.P.'s of AB. AC, and BC. [*Note*—The points A, B, and C should fall by preference to one side of the line. If the V.L. pass between the points, the points will be on two planes, not one.]

(ii) Draw the V.L. of *vertical* planes *perpendicular to P.P.* Form AB, AC, and BC as before. They are lines in a vertical plane. Find their V.P's.

(iii) Draw the V.L. of an *inclined* plane, at any angle to the horizontal, and do with the points as before.

(iv) (*a*) Draw the V.L. of a *directly ascending* plane rising at any angle, and do the same with the points.
(*b*) Draw the V.L. of *directly descending* plane, etc., as before.

(v) Draw the V.L. of a *vertical* plane at any angle to P.P., etc., as before.

(vi) (*a*) Draw the V.L. of a plane *obliquely ascending* towards right, etc., as before.

(*b*) Draw the V.L. of a plane *obliquely descending* towards right, etc., as before.

*In all these cases, if the V.L. is drawn between the points, the points cannot be on the same plane, but will be on two planes.*

LESSON 10. *The position of the Eye in Relation to the Six Planes.*— Make a paper or cardboard model of the relation of the eye to the P.P. and to the V.L.'s of all possible kinds of planes. (See pp. 62, 63, 108 to 112.)

Make the following geometrical drawing: Mark C.V., draw horizon, place eye when 12 feet before P.P. Scale ½ inch to 1 foot. Draw V.L. of vertical planes, perpendicular to P.P., and find its eye. Draw V.L. of planes inclined at 30° to H.P., rising towards left, and find its eye. Draw V.L. of vertical planes making 40° with P.P. towards right, and find its eye. Draw V.L. of planes ascending directly from spectator at angle of 50° with H.P., and find its eye. Draw V.L. of planes obliquely ascending towards right, having their intersections with H.P. at angle of 35° with P.P. towards left, and their angle of inclination with H.P. 40°, and find its eye.

Find the C.V.L. in all cases. The C.V.L. is the end upon the P.P. of the C.V.R. (Central Visual Ray). (Reference, pp. 110 to 112.)

LESSON 11. Height, 5 feet. Distance, 12 feet. Work 6 problems like those in Figs. 85 to 90 taking the positions of the planes from the last lesson. Let the Picture-Lines in all cases be 5 feet from the V.L.'s. Get, in each case, a rectangle, as is done in Figs. 85 to 90, by using 3 feet and 6 feet to right and left of the middle of the P.L., and using 40° on right and 50° on left for the V.P.'s. Get the M.P.'s and measure the sides of the rectangle. [Page 125, line 3—for *square*, read *rectangle*.]

LESSON 12. Height, 5 feet. Distance, 12 feet. Work 6 problems like those in Figs. 91 to 96, making each square 2 feet. Take 9 or 16 squares as one large square and within it draw a circle (Fig. 150). The positions of the planes to be the same as in Lesson 11.

LESSON 13. The conditions being the same as in Lesson 11, find in each of the 6 planes (6 separate problems will be desirable) a square, 6 feet side, of which one side is paralled to the P.P. (as in the last lesson), and the nearest corner on a line which touches P.L. 4 feet left, and proceeds to C.V. or C.V.L., the nearest corner being 4 feet up that line. (In the case of the first problem, dealing with horizontal planes, the nearest corner will be 4 feet left and 4 feet within the picture, but in some of the cases the actual distance will not be 4 feet, owing to the plane being at an angle to the picture). Convert the squares into cubes, as is done in Figs. 97 to 102. This is a lesson on finding perpendiculars to the several planes—a most important matter.

LESSON 14. (i) Distance, 12 feet. Scale, ½ inch = 1 foot. Find a point, A, in the P.P., 5 feet left and 3 feet below level of eye. Join A to C.V. Find a point, B, 6 feet along the line. The line ACV, is the intersection of three planes—a horizontal, a vertical, and one inclined at 45° to the horizontal and descending from left to right. All these planes are perpendicular to the P.P. At B draw 3 lines, one in each plane, and all parallel to the P.P., and each 2 feet long. Mark the extremities C, D, and E. Join CA, DA, and EA, and find their V.P.'s in the V.L.'s of the three planes. Find their M.P.'s, and find their correct lengths. (References, f. 63, f. 64, p. 99, f. 65, f. 81, f. 88, f. 91 to 93).

(ii) Height, 5 feet. Distance, 12 feet. Scale, ¼ inch = 1 foot. A point on G.P. is 6 feet left and 6 feet within. It is the nearest corner of a house of simplest form. The front of the house is 20 feet long and 9 feet high to the eaves. The front is in a vertical plane, making 40° with P.P. towards right. The side of the house is 12 feet wide and becomes the gable end of the house. The slopes of the gable are 45° with the horizontal. Add a door, windows, a chimney, and some railings parallel to the front of the house. Note the V.L.'s of all the planes of the house, and find all the P.L.'s. (References, f. 89, f. 94, f. 103, pp. 119 to 121).

LESSON 15. Height, 5 feet. Distance, 12 feet. A small paved court is six-sided. Five of the sides are equal, 4 feet each, and form five sides of a regular octagon. The six corners of the court, reading from the left, are A, B, C, D, E, and F. The sixth side, AF, of the court touches the G.L., A being 6 feet on left. Each of the five sides AB, BC, CD, DE, and EF, is the base-line of a plane sloping down and touching the G.P. AB is the base-line of an *inclined* plane 30° with H.P. rising towards left. BC is base of an *obliquely inclined* plane rising towards left at the same angle to H.P. CD is base of a *directly ascending* plane also at 30° to H.P. DE is base of a *vertical* plane, and so is EF. Find the V.L. and P.L. for each plane.

Mark the centre of each base line, that is, find points midway between A and B, B and C, etc., and mark them *m*. Run a line up each plane from *m*, perpendicular to the base-line. Upon these lines find points *n*, distant from *m*, in the case of the inclined planes, 2 feet, and in the case of the vertical planes 6 feet and 9 feet. At *n* in each case draw a line *no*, 9 feet long, perpendicular to the plane it arises from. The line *no* is the axis of an object consisting of a pyramid whose base is 2 feet square, and whose altitude is 4 feet, and whose axis is prolonged 5 feet to *o* by a rod, which at its uppermost foot carries a flag 1 foot square. Each of the five planes carries one of these objects, one corner of the square base being on *nm*, and the axis being in the same vertical plane as *n* and *m*. The flag is in all cases to be in a vertical plane (a little consideration will show that the flag is in the same vertical plane as *n* and *m*.) (References, p. 25, etc., p. 94, etc., f. 65, f. 68, f. 70, p. 108, etc., p. 129, etc.).

LESSON 16.—The roof of a building is 78 feet long and 25 feet wide. Its pitch is 50°. The nearest corner, A, is 10 feet left, 30 feet within the picture, and 30 feet below the eye. One of the long sides passes away to the right at 50° to P.P. Along this side are six dormers, each 8 feet wide, rising 4 feet above the eaves, and surmounted by gable roofs of 50° pitch. The dormers are separated by 2-foot spaces, and the outermost are 10 feet from the ends. Draw in perspective, the eye being 12 feet from the picture. Scale, ½ inch = 1 foot.

[As 30 feet is a long distance for the eye above the H.P., take 10 feet first, and, having found the nearest corner, drop it down thrice the distance below the horizon. Then, through the point A thus found, draw a Picture Line, and use the scale which results from assuming the distance from A to horizon to be 30 feet. Use throughout V.P.'s in vertical and oblique planes as much as possible.]

LESSON 17.—Height 5 feet. Distance 12 feet. Scale, ½ inch = 1 foot. A cube stands in the position shown in Fig. 115. Its edges are 5 feet. Two sides are vertical, with one diagonal vertical. The nearest corner of the edge upon the ground is 4 feet right, 5 feet within. The edge vanishes to left at 40°.

Find a point A, 4 feet left, 4 feet within. The exercise is upon dealing with different planes cutting through the cube, as if the cube were embedded in them. Take them in order, repeating the whole exercise if necessary.

(i) A horizontal plane, 2 feet above the ground, cuts through the cube. Find the intersection.

(ii) An inclined plane, whose intersection with G.P. includes point A, and whose inclination to G.P. is 30° towards right.

(iii) An obliquely inclined plane, inclined at 30° to G.P. towards right, and whose intersection with G.P. is at 45° to left, and contains A.

(iv) A directly ascending plane, at 30° to G.P., commencing at A.

(v) A vertical plane, whose trace on G.P. includes A, and which passes through the centre of the cube.

(In such problems, it is wisest to use oblique perspective as much as possible. Even to get the cube, a truer cube will be got by the oblique methods. The interpenetration of one plane with another can be found by the ordinary roundabout methods of finding the intersections of the plane with imaginary verticals from the corners of the cube. But the intersections will be best managed by oblique methods. *Where the V.L.'s of two sets of planes cross is the V.P. of all intersections of those planes.* Thus the V.L. of the inclined plane passes through the horizon at C.V. C.V. is therefore the V.P. of all intersections of all inclined planes at the same angle with all horizontal planes.

As a rule, one commences with a vertical plane. Vertical planes are indeed valuable means of getting from one plane to another.)

LESSON 18.—The previous lesson provides us with subjects sufficient for the study of drawing shadows. We have the cube standing on its edge on the G.P., and we have the same cube partly embedded in other planes. Our only upright lines are the two vertical diagonals. These are not actually lines of the cube, but can stand as such and serve our purpose. Indeed, by casting the shadows of these verticals, we can prove whether our shadows of the slanting lines are true.

(A) *Shadows from artificial light—* [*Note*—If AVP$_1$, Fig. 115, were below a certain position, the RS$_1$ would be on the right of VP$_1$, *above* the horizon, but still on the line through *seat* and VP$_1$. When *Light* and AVP are on a line parallel to the line through seat and VP, no RS is found, and the shadow is parallel to these lines. When Light and AVP coincide, the point is also RS.]

Find a point 4 ft. left and 6 ft. within. At this point erect a line 8 ft. high. The summit is an artificial light.

(i.) Cast shadows from the cube on the G.P. (Fig. 115).

(ii.) Find the inclined plane as in Lesson 17, and cast the shadow of the cube upon it.

(iii) Find the obliquely inclined plane as in Lesson 17, and cast the shadow of the cube upon it.

(iv) Find the directly ascending plane, as in Lesson 17, and cast the shadow of the cube upon it.

(v) A vertical plane, perpendicular to the picture, touches the corner of the cube which is farthest on the right. Find the shadow of the cube upon it, considering the ground-plane non-existent.

(vi) A vertical plane parallel to the P.P. is behind the cube which touches it by its hindmost corner. Find the shadow of the cube upon the plane, the G.P. being considered non-existent.

(B) *Shadows cast by the Sun—*

(vii) Repeat all the problems, considering the sun to be before the spectator, its rays inclined at 40° to the ground and in planes 50°, with P.P. towards left. Problem (vi) is unsuitable for this lighting. [*Note*—Shadows on planes parallel to P.P. are parallel to lines joining *Sun* and V.P.]

(viii) Repeat all the problems, considering the sun behind the spectator or left, its rays inclined at 30° to G.P., and in planes receding towards right at 40° to P.P.

(ix) Repeat all the problems, considering the sun in the plane of the picture, its rays at 45° to the G.P.

[In all these problems the main subject should be worked in strong line, and the shadows worked on tracing-paper.]

# PERSPECTIVE FOR ART STUDENTS

## 1.

## What Perspective is.

WE all know that when we draw a view of an object we have to make its more distant parts smaller than its nearer. We all know that as things recede into the distance they become smaller in appearance. In short,

FIG. 1.—An object diminishing as it recedes.

we know that some such regular diminution of parts as is shown in Fig. 1 must take place.

This regular diminution of receding parts is called *the perspective* of the objects; and we draw *in perspective* when we pay due attention to this diminution.

B

The laws of perspective affect all objects, no matter how varied or curious their form, but the influence of these laws is most evident when the objects are of severe geometrical shape, such as railway-tracks, straight roads, buildings, ard the like.

**2.**

## Cuboid Forms the most suitable for Perspective Problems.

It is no exaggeration to say that the parallelopiped (brick-form) and the cube are the most convenient solids for perspective representation. As a rule all forms are considered to be enclosed in one of these.

Both the cube and the brick-form consist of lines in three directions only. These are lines to the right, to the left, and upwards. They are all perpendicular to one another. In Fig. 2 the influence of these cuboid forms is illustrated. Point A is the corner from which start the three lines—to right, to left, and upward.

He who has mastered these three lines, especially if he can sketch them in by freehand, has practically mastered perspective. Of course the solids do not necessarily occur lying flat on a horizontal surface, such as the ground. He, too, who would master *sketching in perspective*, must certainly be able to draw from imagination a cube in any position, but most assuredly when lying on the ground. He must be able to draw upon the sides of the cube stripes parallel with the edges. It is by this means that all the form is obtained in Fig. 2. It is little use the student drawing cubes and such like from nature. One thing alone will help him, and that is the determination to make his cube, whether he draw from nature or no, look cubical, and practice till his

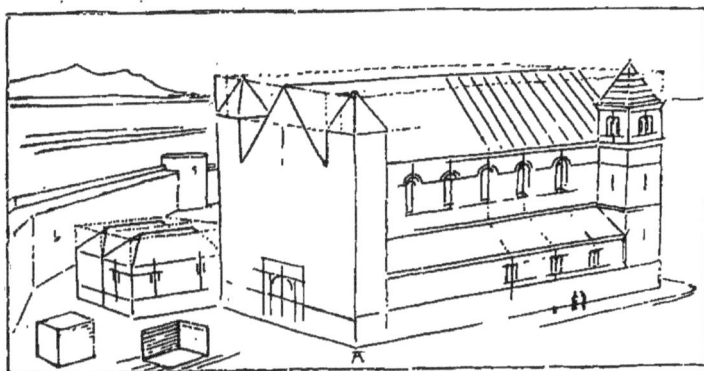

FIG. 2.—Cuboid form prevailing.

hand makes the necessary radiation of the lines automatically.

The student will see how, as he progresses with the subject he gets into the habit of always finding, as practically the first thing to do in every case, how to get his lines to the right, his lines to the left, and his upright lines. Usually he finds he attacks his object best by getting its base on the plane the object stands on, and then raising perpendiculars to the plane for the upright edges. For this the different kinds of planes demand different treatment, as will appear in the proper place.

## 3.

## Why Things vanish.

We see by means of a lens collecting rays of light from the objects visible, and conducting them to a screen (retina) at the back of the eye. The lens and retina are the apparatus of vision. The image thus cast upon the retina is apprehended by the optic nerve. The lens and

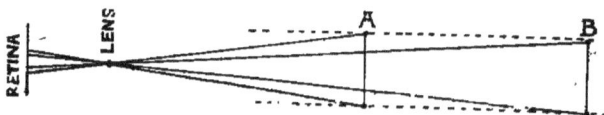

Fig. 3.—Objects of equal height and at different distances from the lens giving images of different size on the retina.

retina are in fixed positions, so that the nearest and the most distant objects must alike be imaged upon the same retina. A simple diagram, Fig. 3, serves to make clear that the rays of light from the more distant object, B, cannot expand to the extent reached by those from A.

## 4.

## The Picture-plane.

The image on the retina of the human eye is no more than half an inch across. A drawing of such dimensions

would be quite useless if obtained. The idea of repro-
ducing the actual image on the eye is therefore
abandoned.

Instead of the rays of light being regarded as piercing
the lens and expanding upon the retina, they are held to
be intercepted *before* they reach the lens, which is hence-
forth called the EYE, by a plane called the PICTURE-
PLANE.

It will be seen by Fig. 4 how that this picture-plane
may be placed at any distance from the eye with no other

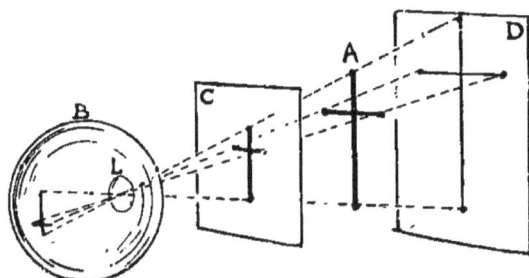

FIG. 4.—A, the object; B, the ball of the eye; L, the lens; C, a
picture-plane between the object and the lens; D, a picture-plane beyond
the object.

result than that of altering the *size* of the image on the
plane. It will also be seen that the image is no longer
upside-down as it was on the retina.

All images on the picture-plane are produced by intersections. Thus in Fig. 5 rays of light are converging
from the actual real points A to the eye. These four
rays severally pierce the picture-plane, and produce there
the points B. Each point signifies to the spectator (eye)
a corresponding point A, for B is precisely in line with
A. As far as the spectator is concerned, B and A are

both one; in fact, the whole ray of light is, when viewed from the eye, merely a point. It is very important that

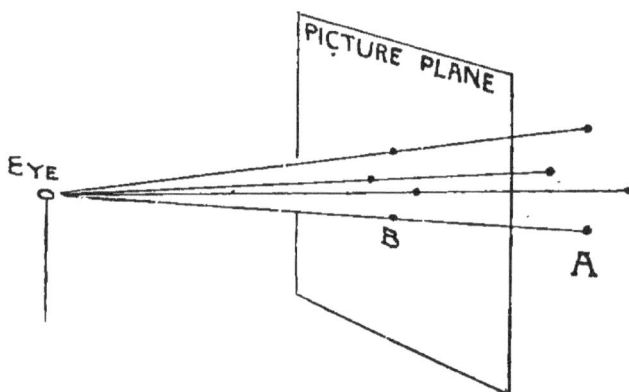

Fɪɢ. 5.—The image, B, produced on the picture-plane by rays intersecting the plane.

the student should understand this, for it is somewhat startling to have to accept a point on the picture-plane,

Fɪɢ 6.—The artist touching a point some miles away.

perhaps not above 12 feet from one, as actually a point miles in the distance. But of course, to the spectator, if a point on the picture-plane *hides* the point

in the far distance, it *is* that point, for it takes its place
in the view.

The relation between picture-plane and spectator is
therefore as follows. The picture-plane is some distance
before the spectator, perhaps 12 feet, but the distance
will vary according to the size required for the drawing,
as has been sufficiently shown by Fig. 4. Innumerable

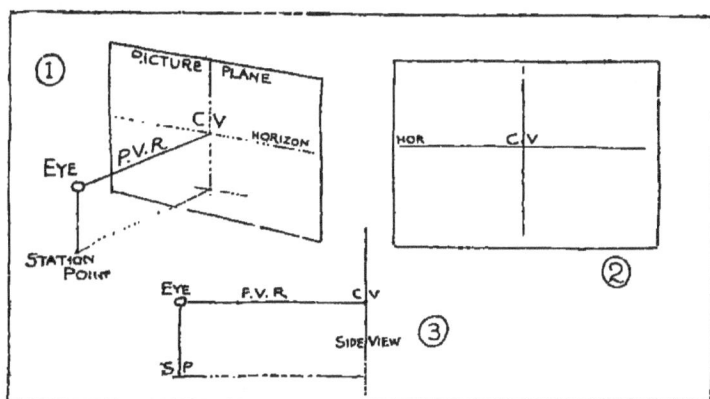

FIG. 7.—The eye and the picture-plane.

rays of light will pierce the picture and converge to the
spectator's eye. Of all these rays the one which occupies
the shortest distance between the picture-plane and the
eye is known as the Principal Visual Ray, P.V.R. This
shortest distance is, of course, the stated distance of the
picture-plane from the eye, say 12 feet.

The point where this P.V.R. pierces the picture-plane
is the Centre of Vision, C.V.

## 5.

## Objects protruding before the Picture-plane.

The image remains true whether the object protrude before the picture-plane or not. For its image is obtained by radial projection from the eye, so that the rays as truly cast the images of objects before the picture *upon* it, as they bring up the images *to* it of objects behind. Precisely similar objects one before the other behind the picture are given in Fig. 55, p. 81.

It is quite unusual for the object to project before the picture-plane in examination questions, but instances are given later on, among the hints for designers and architects, wherein it is an advantage to allow parts of the object to come through.

In Question 5 (April Examination, 1901) solved on p. 205, the object protrudes before the picture-plane.

## 6.

## Sketching and Working Perspective Drawings.

A capable draughtsman can make a satisfactory perspective drawing merely by skilfully radiating the lines in a suitable manner. He must, however, know the main rules of perspective. He must know, that is—

*That parallel planes vanish in the same line;* and

*That parallel lines vanish in the same point.*

These are the two great laws of perspective. He must know the following subsidiary fact :—

*Lines in a plane have their vanishing points on the vanishing line of the same plane.*

Just as the sketcher must guess where his vanishing points and vanishing lines are, so he must guess his measurements. A very great deal can be done in this way with care; but everything depends upon skill, and a good guess where a vanishing point should be will avail one little if the lines supposed to be converging thither go somewhere else instead.

The difference between a sketched and a worked perspective is one of accuracy, of accuracy due to the statement to the draughtsman of certain particulars which he can use mathematically to obtain his points and measurements with absolute precision. To the degree that these particulars are wanting the drawing must be *sketched*, aud cannot be *worked*.

What is the least information we can do with? The least is a general instead of definite knowledge of the direction of the lines and planes. In making our sketch, we, however, convert this general or vague knowledge into precise and exact guesswork. If we know that the lines of a wall vanish somewhere on our right we do not vanish them "somewhere," but as definitely as we can in some direction which we guess to be about right.

## 7.

## Sketching—The Three Lines.

In paragraph 2 we have seen that perspective deals most readily with objects of cuboid form, and that to deal with objects which are not cuboid it is advisable to imagine them enclosed in a rectangular framework. Fig. 2 illustrates this. He who sketches in perspective must master the three lines, the lines in three directions which bound the object.

Suppose, therefore, we approach the representation of a cuboid object such as that in Fig. 8—what have we to consider, what have we to do, to make our drawing perspectively correct? It is the object of this book to answer that and similar questions.

FIG. 8.

Let us exhaust first the knowledge we have which will affect this case.

1. We know our object has six sides, of which we are to see three. The uppermost surface, or top, is similar to the base, or side, on which the object stands. The base is represented in our

drawing at present only by the two edges nearest us, the two lowest lines of the drawing.

2. We know that it is the nature of a cuboid form to have the base and top precisely similar. We know that not only has the top four edges, of which those opposite one another are parallel, but we know that the base repeats these conditions.

3. The top and base are thus so similar, and so placed in relation to one another, that they must be always thought of together. Together they give us eight lines (Fig. 9), of which four pass to the right, and four to the left. Of each

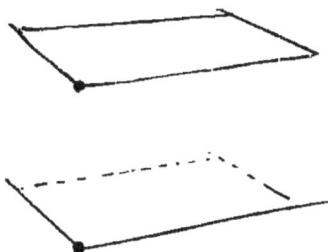

Fig. 9.—The top and base of a cuboid object.

of these two sets one will be invisible in the complete object.

4. We know already that the diminution of objects receding into the distance is regular and subject to law, and that we express this regularity and this law by a reference to parallel lines—we say that they vanish in the same point. All forms and lines, no matter how irregular, are subject to diminution according to distance, but the diminution is not expressible in words, except in such an instance as that of parallel lines or planes.

We know, then, that four of our eight lines in Fig. 9 will vanish to a point on the right, and that the other four will vanish to another point on the left.

This fact, which should be so self-evident, is often overlooked by students, as all teachers of perspective will admit. Students remember to vanish the front lines properly, and then straightway vanish the back lines anywhere! No perspective can be done without understanding, and the obvious fact here pointed out is the very first step. If it is properly taken, there is not much difficulty with the rest.

5. We know. also, that what is true of our top and base is true also of each of the two pairs of sides. The front and back correspond exactly. There are the eight lines as before. Four pass to the right, and four pass downward (Fig. 10).

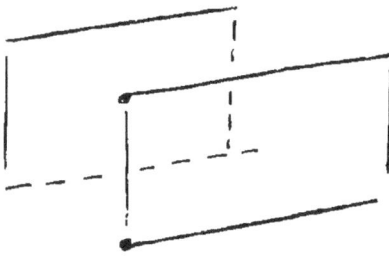

Fig. 10.—Two of the sides of a cuboid object.

6. We know that "parallel lines vanish in the same point," but we must add to this that parallel lines which are also parallel to the spectator (the expression used in perspective is *parallel to the picture*) are not regarded as receding, and so do not vanish. Hence our four descending lines in Fig. 10, if supposed to be vertical in fact, are drawn vertically without any convergence.

There is no need to separately illustrate and describe

the remaining pair of sides, which correspond in every way, both of fact and treatment, to the front and back.

The difficulties are thus centred in the three lines expressed in solid line in Fig. 11. We may take the central point at the base or at the top as we please, and as is illustrated in the figure, but each line must govern three others. If, then, we

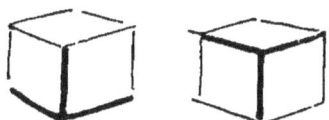

FIG. 11.—The three lines.

know where each of these three lines goes, we know all*
we need to know.

## 8.

## The Three Lines—Their Directions.

The three lines indicate by their directions the position of the cuboid object. To the sketcher the position is what it appears to be, because the final judge in these matters is the eye. It is no use saying an object is lying flat on the ground if it looks tilted, nor saying its sides are equal if they do not look square. The *worker* of perspective, on the other hand, defends himself by showing that he has followed the proper rules, and if the result is wrong it is because perspective as a science is faulty. There is, no doubt, sometimes a discrepancy between the result which satisfies a worker and that which satisfies a sketcher. The fact that perspective, when worked by rules, does not permit any convergence of line parallel to the picture, is at the root of the discrepancies. We shall see later on where the two systems clash.

The variations of the directions of the three lines are due to—

1. The distance of the object from the spectator.
2. The distance of it on his right or left.
3. The distance of the object below or above the eye.
4. The angular position of the object.
5. The position or tilt of the plane on which the object stands.

Each of these conditions will have to be treated separately.

# 9.

## The Distance of the Object from the Spectator; how Aspect is affected by it.

At a later stage, when the working of perspective is explained, the student will find he has to recognize a front and a back edge to the ground. He will see that this *front* edge is very arbitrarily chosen. There is really no front edge to the ground in the sense that there is a back or distant edge. The back edge is the horizon, with which all are conversant who have seen the sea. If we like we can take the position of the spectator's feet as the beginning of the ground. But no one's picture ever comes so low down as to take in his own feet, so we may leave the matter in its natural vague state.

What, however, is important to notice is that from the front edge of the ground, wherever that may be, to the back edge is an immense distance, expressed on a drawing by a very narrow space. The higher a point is, therefore, assuming it to be on the ground, up toward the horizon, the further away from the spectator it is.

In Fig. 12 we have several squares lying on the ground-plane. It will be seen at once that the further away the square is, the less its lines are inclined. The

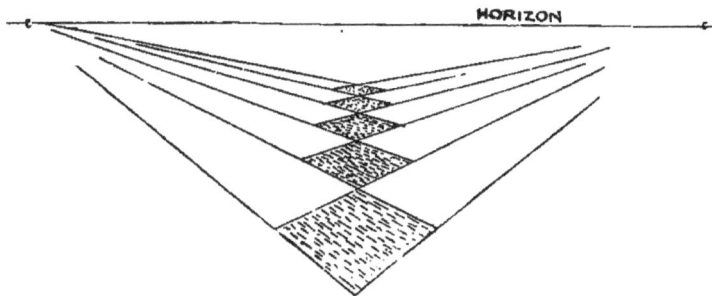

Fig. 12.—Five squares of equal size placed successively at greater distances from the spectator.

square becomes in the drawing a very sharp diamond, with its two front edges making so large an angle as to be very little removed from the straight.

The two front edges of a square may thus be an indication of the distance the object is from the spectator.

Hence in Figs. 13 and 14 the difference of pitch of the
two front lines, whether we take those made by the edge

FIG. 13.—A building at a considerable distance before the spectator.

of the building on the ground or the edge of the
building against the sky, indicates a difference of

FIG. 14.—A building nearer the spectator than that shown in Fig. 13.

distance, Fig. 13 representing a greater distance than
Fig. 14.

Another notable effect of the difference of distance is
seen in the relative heights on the drawing of vertical
lines of equal height actually. The difference is greater

when the distance is less. In Fig. 13 the columns on the building differ very little in height, in Fig. 14 considerably. Subjects arranged at a considerable distance have, therefore, an equality of size pervading the parts which gives to the work a monumental character, and fits it better for decorative purposes. I have already referred to this in "Figure-drawing and Composition." We are thus able to treat a considerable area of ground space without greatly changing the heights of our figures or buildings.

On the other hand, a nearer view with its short range, and consequently boldly tilted lines, has a realistic vigour which is sometimes very attractive.

### 10.

## The Distance of Objects on the Right or Left of the Spectator; how Aspect is affected by it.

Students of solid geometry (and none should attempt perspective who have not some knowledge of that science) know that the position of an object is described according to its relation to two planes, one horizontal, the other vertical, and together forming a corner such as is made by a half-open book. Where the two planes pass into or through one another is the intersection between them, and is commonly figured as "XY." If one attempts to describe the position occupied by a

C

thing, one has always to adopt some such bases of position as these planes and their intersection afford. If the student who knows nothing of solid geometry will take a circle of paper and a square of paper, and upon each place a mark, anywhere, and will then try to describe the position of the mark, he will find it impossible in the case of the circle, unless the mark occurs absolutely in the centre, and even then, if the mark be, say, a square, he will not be able to describe its angular position. One cannot hold a circle the right way up, or look at it from the "front." No such terms apply to it. But the square paper immediately provides bases from which measurements can be taken. If our mark is 2 inches from one side and 3 inches from one of the sides adjoining the first side, there can be no doubt as to position of the point.

In perspective (as an art or science) we describe positions in the same way. Thus in Fig. 15, AB and CD are two squares or cubes placed before the spectator in the position seen. The plan of the picture-plane is given as a line, and it is this line which is used as the basis for the description. AB and CD are both the same distance beyond the line, and would be described as so far "beyond the picture-plane," or "within the picture."

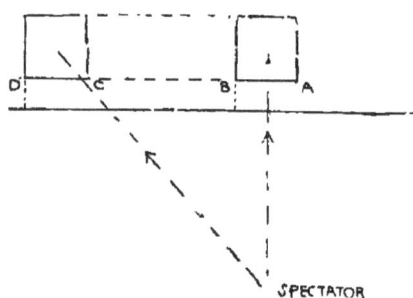

FIG. 15.—Plan of two objects, the picture-plane, and the spectator.

CD is certainly much further from the spectator than AB, but this is overlooked in the rules by which perspectives are worked, because to take cognizance of it would involve us in tasks which would not have a result proportionate to their difficulty.

CD is to the left of the spectator. How does this lateral placing of the object affect the appearance? The perspective result according to rule will be that given in Fig. 16. There we have two squares, A and B.

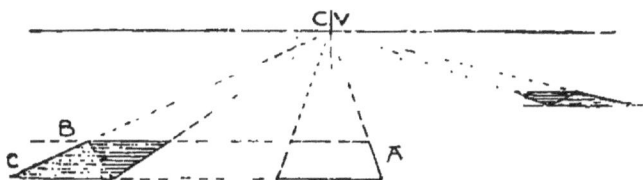

FIG. 16.—Squares on the left and on the right of the spectator.

The square B is supposed to have the same angular position toward the picture-plane as A; that is, in this case, they are both parallel to it. The student will not fail to notice that the square B looks a little "out." The corner C particularly protrudes too far, and the half of the square which is lightly shaded looks too large to be a proper complement in a square to the darker half.

Nevertheless square B is correctly drawn according to the rules of perspective. It is in what is called parallel perspective—of which more is said later on. This square illustrates, then, the great defect of parallel perspective; which gets so wrong when far from the centre of the picture. It was, however, very largely used in old times, and it is interesting to see how the old masters avoided the over-extended angle referred to in Fig. 16.

Here, for instance, in Fig. 17 is a very common form
which design took when perspective was permitted to
control it.  On either side of the room are three pedestals
with the bases and part of the shafts of columns above
them.  It will be seen that the objectionable angle is

FIG. 17.—St. Matthew, from a sixteenth-century woodcut.

cleverly hidden in each case.  Undoubtedly, in examples
such as this, a certain monumental or decorative dignity
is sought and obtained by keeping to the parallel arrangement of the lines, and from a " perspective " point of view
this is the sound thing to do.

Draughtsmen and artists who are more concerned with
an ordinary possible-looking drawing, would, if sketching
their perspective and not working it, assume the cubes at

the sides as being at a slight angle. They perhaps would not think about it, but would unconsciously lift the level line a little. They would assume, that is, that the actual facts of position as indicated by the dotted arrows of Fig. 15 indicate that the cube CD in that diagram is really rather anglewise toward the spectator, and should be treated accordingly. They would draw the form somewhat like the square on the right of Fig. 18, where the

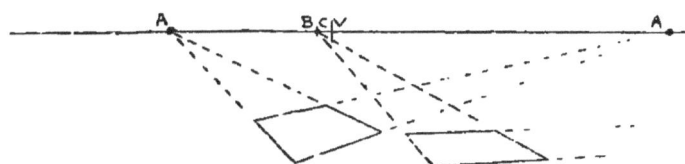

FIG. 18.—Squares at angles to the picture. That vanishing to B is only just off the parallel.

square which vanished to B is only just off being parallel to the picture. Were it parallel to the picture, it would have two lines vanishing to CV, and the other two drawn level. If it vanish ever so slightly to one side of CV, say to B, its formerly parallel edges will rise, and will meet the horizon some long way off on the right. The further the one V.P. gets to the left, the more the other creeps up from the right toward the centre.

An instance of the " easing " of this harsh effect caused by the objects at the sides being kept in strict perspective is given in Fig. 19.

The incongruity referred to has been treated before, and sometimes systematically. The *Art Journal* for 1849 contains an exposition of the advantages of *curvilinear* perspective. It is so much a matter, however, for the eye

and hand of the artist, and so little a matter for the
draughtsman, who wants perspective as an assistance

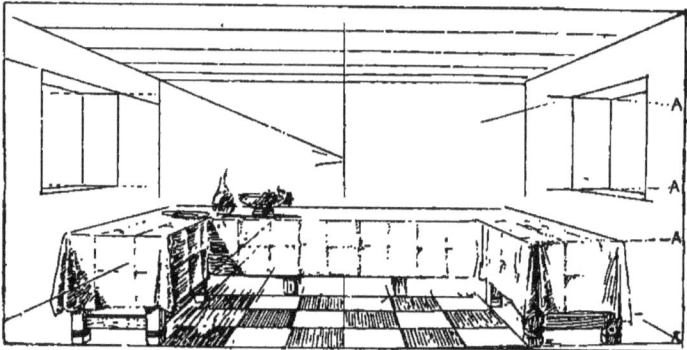

FIG 19.—A table and window.  On the left the lines are left parallel,
on the right they are "eased" a little.  The lines marked A are
converging.

toward a result largely "safe," if less attractive, that
enough has probably been said of it.

In Fig. 20 both cubes are at the same angle to the
picture, 45°.  That on the left has a necessarily altered
form, and the student will notice that as FD becomes

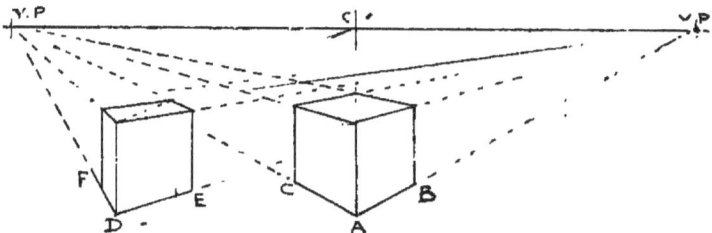

FIG. 20.—The different aspect of objects placed in different lateral
positions.

nearer and nearer upright, DE becomes nearer and nearer
level.  Various positions across the picture would therefore illustrate this reciprocal change of inclination.

## 11.

## Different Positions below or above the Eye; how they affect the Aspect.

The student will hardly need to be told what change will take place if the object be removed to a greater depth below or to a greater height above the level of the eye. At the level of the eye any form—square, circle, or what not—will appear as a line merely. All its edges will be reduced to the same direction, its surfaces will be lost entirely. We do not any of us need to be told that if we hold a piece of paper or cardboard edgeways towards us, we see only the edge of it. As we lower it or raise it we see on to or up to one of its sides, and the further it is from the level of the eye the more its appearance approaches its actual shape.

What needs, however, to be impressed is that the two front edges (to keep to our example)—such edges as are shown in Fig. 11, p. 13—become more angular as they represent forms lower and lower down below the level of the eye, or forms higher and higher above it. The change is evident enough in the cubes in Fig. 20, and it is further illustrated in Fig. 21. As the objects fall lower and lower or rise higher and higher, the angles become more obvious, while, contrariwise, the nearer the level of the eye they are the more likely to be almost continuous, or so delicately angular as to be very little removed from the level.

When the front edges, then, are only delicately angular
to one another, we infer that either the object is (as

Fig. 21.—Different tilts of lines below the eye at different depths.

according to this paragraph) very little below or above
the level of the eye, or (as according to paragraph 9) the
object is at considerable distance from the spectator.

Simple as the facts are that are brought out in these
paragraphs, it is remarkable how rarely a student has
any difficulty with perspective who masters these easy
and obvious principles.

**12.**

## The Difference in Angular Position of the Object; how it affects Aspect.

There is no need to deal at length with this one of the
five influences which affect the aspect of objects. In
Fig. 18 we have an example of different angular

positions. We noted in that case the *reciprocal* vanishing points, and how a movement of one of them away from the centre drew its fellow further in.

In Fig. 20 we have two cubes in the same angular position, but giving very different aspects. We can see at once that the aspect FDE could occur where CAB does if the cube FDE were in a new angular position.

We need only remark further that care must always be taken that neither of the outer points of our two front edges, as point C and B in Fig. 20, must be ever lower than the middle point, such as A, if the object is supposed to be lying on a horizontal plane.

### 13.

## Differences in the Inclination of the Plane on which the Object stands.

So far we have been supposing that our object was lying on the ground-plane; its upright lines were still vertical, and were not tilted over as they would be if the plane of support were itself suddenly inclined. The ground-plane is always supposed to be level. Any inequalities, as hills, are unevennesses of the ground, and do not alter the ground-plane—for a *plane* cannot be uneven.

Objects in perspective are always held to be standing on or lying in planes. Obviously, therefore, a change in the tilt or inclination of the plane will make the object heel over. What result has this tilting on the aspect of the objects?

Now, there are *seven* possible *kinds* of tilts or inclinations, and therefore, to see how the objects are affected, we shall have to pass in review all these seven possibilities.

We shall have to give names to these seven positions; and having done so we shall find that one can be dismissed as of little moment (this is the position parallel to the picture), while three only will cause any great amount of trouble.

We have already seen that positions are described in perspective, as in solid geometry, by the relation the object bears to two assumed planes—the horizontal and the vertical. In perspective these two are the ground-plane and the picture-plane.

The planes are described according to the angles they make with these two planes. They are all illustrated in Fig. 22. Thus A is a *vertical* plane, perpendicular to the picture-plane; B is a *vertical* plane making an *angle* with the picture-plane (50° towards the left). C is a second instance of the same position of plane; it is at 40° to the right. D is a *horizontal* plane parallel to the ground-plane and perpendicular to the picture-plane. E is a *directly ascending* plane at 35° to the ground, F is a second instance of the same position. G is an inclined plane whose trace on the ground is perpendicular to the picture. In this matter it resembles A, one of the vertical planes. G is, however, inclined to the ground, and its intersection with the picture-plane shows an

angle of, in this case, 45° with the ground-line. H is an
*obliquely ascending* plane. Its companion, an obliquely
descending plane, is not shown. Either resemble vertical
plane C in one particular—the trace on ground-plane is
at an angle to the picture-plane. In this case it is 30°.
This plane also makes an angle with the ground-plane.

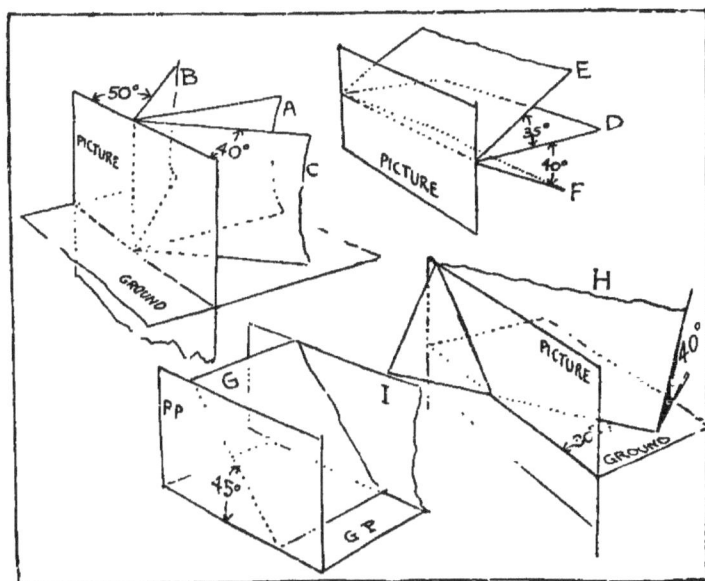

Fig. 22.—The seven positions of planes.

This is shown in black, and is, say, 40°. The oblique
plane differs from all the others in having a trace on the
picture-plane unlike either of its "edges." The plane
projects through the picture-plane, and the junction
between the two is an oblique line.

The seventh position is that parallel to the picture and
perpendicular to the ground. None of the lines in it

will vanish or converge to their fellows. It is marked I
in Fig. 22.

Perhaps the easiest classification of the seven positions
is as follows :—

1. Parallel to the picture (I).

Planes perpen- ⎧ 2. Horizontal (D).
dicular to the ⎨ 3. Vertical (A).
picture ⎩ 4. Inclined (G).

Planes inclined ⎧ 5. Directly ascending or descending (E or F).
to the picture ⎨ 6. Vertical (B or C).
⎩ 7. Obliquely ascending or descending (H).

The chief difference between these several planes is
shown by the *perpendiculars* to them. The perpendiculars
vanish as follows :—

Of No. 1, to centre of picture, C.V.

Of Nos. 2, 3, and 4, do not vanish, but remain parallel
to one another and to the picture-plane.

Of Nos. 5, 6, and 7, vanish in the contrary direction to
the plane.

Putting this fact another way, the perpendiculars to
planes Nos. 2, 3, and 4 do not expand or diminish the
object they bound. Those of planes Nos. 5, 6, and 7 do,
as also do those of No. 1.

## 14.

## Different Inclinations of Planes; Effect on the Aspect; Planes Perpendicular to the Picture.

The planes perpendicular to the picture have been numbered 2, 3, and 4. They are further illustrated in Fig. 23.

In these diagrams, the inclined and the vertical are shown as turned up as by a hinge from the horizontal.

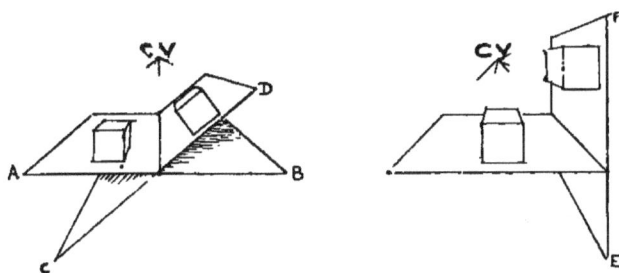

FIG. 23.—The planes perpendicular to the picture. AB = horizonal, CD = inclined, EF = vertical.

It will be seen that the change in the aspect is trivial, chiefly because the upright lines remain *perpendicular* to the front edge of the plane. The front edges here are AB, CD, and EF.

### 15.

## Different Inclinations of Planes; Effect on the Aspect; Planes inclined to the Picture.

These planes are the *directly ascending*, the *vertical* when inclined to the picture, and the *obliquely* ascending.

The *directly ascending* is like a horizontal plane hinged upon its front edge. In No. 1, Fig. 24, a cube is shown standing on a horizontal plane; its upright edges are quite vertical, and its receding sides vanish to C.V. If now we tilt the plane up so that the vanishing point (C.V.) rises higher, to C.V.L., we convert the plane of support into one directly ascending. What were upright lines will now vanish to a point somewhere under C.V. All perpendiculars to a directly ascending plane vanish somewhere under C.V. The new cube is shown in dotted lines.

The *vertical* plane (No. 2) at an angle to the picture is like a vertical plane perpendicular to the picture hinged on its front edge. Hence the V.P. shifts along, and the lines projecting from the plane no longer are drawn with the T-square, but vanish to a V.P. somewhere on the other side of the drawing. All these V.P.'s are on the horizon in this case, and the *V.P.*'s of perpendiculars to all vertical planes inclined to the picture *are always* on the horizon.

The *oblique plane* differs from the directly ascending in having its intersection with the ground at an angle

with the picture. In No. 1 the ascending plane is
hinged up at the lower front edge of the cube. That

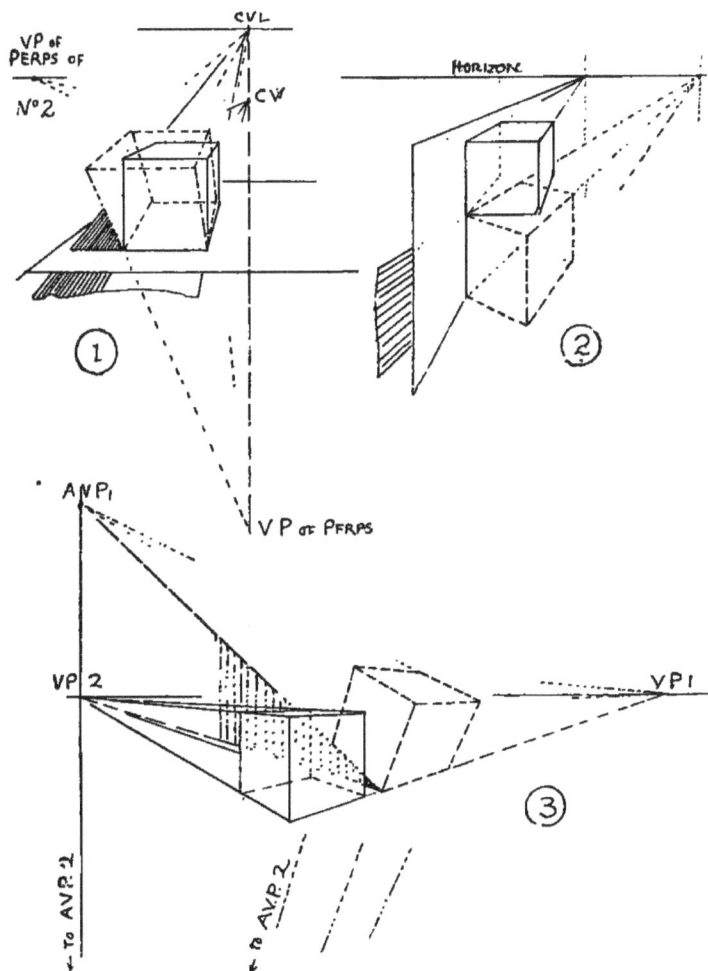

FIG. 24.—Objects in planes inclined to the picture.

edge, then, is parallel to the picture (to the ground-line
in the diagram). In No. 3 the cube standing on the
horizontal plane is placed angularly, and vanishes to

right and to left. By raising V.P.$_2$ up, we tilt the
cube forward into an oblique plane. The perpendiculars
vanish to A.V.P.$_2$, somewhere below V.P.$_2$.


## 16.

## Vanishing Points of Lines in the Various Planes.

We now know that the lines of our object will be
inclined according to the stress of the five conditions
enumerated on p. 14, and elucidated in the preceding
paragraphs. We know that the lines which are parallel
will vanish somewhere, sometimes high up, and sometimes low down, on the paper. We have now to get
to know more precisely where these *vanishing points*
will be, and what conditions will follow in the various
cases.


### (a) Horizontal Planes.

Fig. 25 illustrates the vanishing of lines in horizontal
planes. The object is a cubical solid, with a projection
above. It was first sketched in in the rough. This
rough drawing gave the vanishing much as it now is.
It was necessary to determine definitely where line AB
would meet line D. This convergence gave a vanishing point, V.P.$_1$. As all the lines (except the verticals)
were assumed to be in horizontal planes, it was certain

that V.P.$_1$ was a point in the vanishing line,* or back edge, of all horizontal planes, that is, in the horizon. The horizon was thus drawn level through V.P.$_1$, and upon it by guess was found V.P.$_2$, the vanishing point on the right. It was then possible to make the figure precise. The height of the-vertical on A being given as right, AC was required to be of the same length. The front of the object is therefore square, and to determine C in its proper place, AC was guessed equal to the

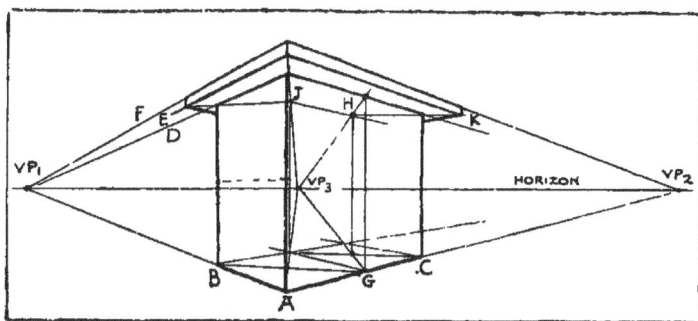

Fig. 25.—Lines in horizontal planes.

vertical on A—or, in short, the square was made to look square. AB was required to be in length half of the height of the front, so a square on half the height of the vertical on A was guessed.

The projection above was obtained by the use of the diagonals. Part of the practice of perspective is certainly the invention of ways and means, and the alert

* The term " vanishing line " does not mean a line which vanishes but the line in the distance in which a plane appears to vanish. See Paragraph 20.

D

application of labour-and-line-saving methods. A projection carried equally about an angle has the junction of its two sides on a line bisecting the angle. In joinery this line is the *mitre*. If, then, we get the direction of the mitre, we can poke out our corners easily. The front AC being twice AB, we guess G, so that AG looks equal to GC. We have thus our base divided into two squares. The squares being already complete, we have only to join their corners, and by so doing we obtain the diagonals, which are in fact the mitre-lines. Where they cross is the centre of the square. As we have these in two squares, and shall need them above, a vanishing point for them will be useful. Draw, therefore, the diagonal from G up till it cuts the horizon. This gives us V.P.$_3$, the vanishing point of the diagonal dividing the angle BAC. To gain the centres of the squares above, we raise a perpendicular above G, and, having struck the top line, run back to V.P.$_3$. By throwing a line up from the centre of the square below we obtain H, the centre of the farther half of the top of our object. We obtain the other centre J in a similar manner, and by drawing lines from these centres out through the top corners of our object, are able to push out our projection as far as we wish. In this case the extent of the projection was not calculated.

This example shows that parallel lines, although not on the same planes, meet at the same point.

Note that the perpendiculars to horizontal planes are drawn vertical without vanishing.

#### (b) *Vertical planes.*

In our last subject the front and side of the object are in vertical planes—planes like walls receding, the one to the V.P. on the left, the other to the V.P. on the right. Any straight line, taken by itself, no matter what its position, or what other plane or planes it may be in, can also be in a vertical plane. This fact is taken advantage of largely in the more advanced stages of worked perspective.

Suppose our subject be a pediment (Fig. 26), that triangular low gable so largely used in classic architecture.

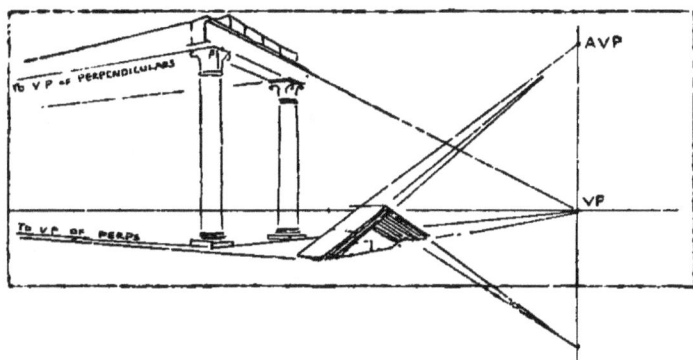

Fig. 26.—Lines in vertical planes.

Suppose it to lie on the ground in the simplest position, that is, with its sloping edges all at the same angle to the ground, and in vertical planes. The pediment does not, therefore, nod forward at all. Where will its sloping edges vanish ? If we were to take the two sloping lines bounding the top, towards us, as right, we could join them and so get the vanishing point A.V.P. (accidental vanishing point). But the edges are not to be relied upon. We must therefore find the vanishing point by other means. The only other means is to find where the vanishing line cuts the horizon, or, in other words, find horizontal lines also in this vertical plane. Suppose the two columns with their fragment of entablature be in the same plane, or rather in planes parallel to those containing the pediment. All the planes will vanish in the same vanishing line. Now, the entablature and the bases of the columns provide us with horizontal lines in our vertical planes; these horizontal lines will vanish both on the horizon and on the vanishing line of their vertical planes. As they can have but one vanishing point, it is clear that that point must be in the two planes. V.P. is the point, so the horizon is drawn level through it, and the vanishing line of the vertical planes vertically through it. Assuming now that one line of the sloping pediment be correct, we prolong it to this vanishing line, where we obtain the accidental vanishing point, A.V.P. The other slope of the pediment being at the same angle to the ground, we shall find its V.P. as far below the horizon as A.V.P. is above it.

The term *accidental vanishing point* has been used. Some draughtsmen call vanishing points accidental when they do not occur on the horizon. The only reason for this can be clearness, since all vanishing points are precisely alike in scope and use.

Note that perpendiculars to the vertical planes, which in this example are slanting off to the right and do not run straight into the picture, will vanish to a point *on the horizon* somewhere out on the left.

(c) *Vertical planes perpendicular to the picture.*

It has been several times impressed upon the reader that perspective prefers dealing with cubical forms, forms which include the three directions of solid measurement. The reader will grasp at once that if two of these directions—as, say, two contiguous sides of a square—be *in* a plane, the third direction must project out from it perpendicularly.

The difference between vertical planes which make an angle with the picture and those which are perpendicular to it, is seen in the treatment of the perpendiculars to the plane itself. In Fig. 26 many of the lines are converging

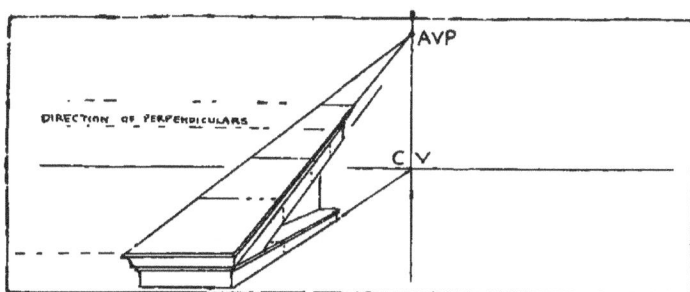

FIG. 27.— Lines in vertical planes perpendicular to the picture.

to vanishing points on the vanishing line of the vertical plane on the right. The lines perpendicular to these are seen receding to the left, and apparently would meet the horizon some distance out of the drawing.

In the case of the vertical planes perpendicular to the picture, the perpendiculars to them are of necessity in

planes *parallel to the picture,* and are therefore themselves parallel to the picture, and will not vanish. Hence
in Fig. 27 the lines going to the left, if such a term can
be used, are all drawn parallel to the horizon—level, that
is, and it would not matter how far to right or left the
object occurred, one set of its lines would always be
parallel and horizontal.

(*d*) *Inclined planes perpendicular to the picture.*

These planes are very similar to the last in having
the perpendiculars to them not vanishing, but drawn
perpendicular to the vanishing line. In Fig. 28 A, B, and

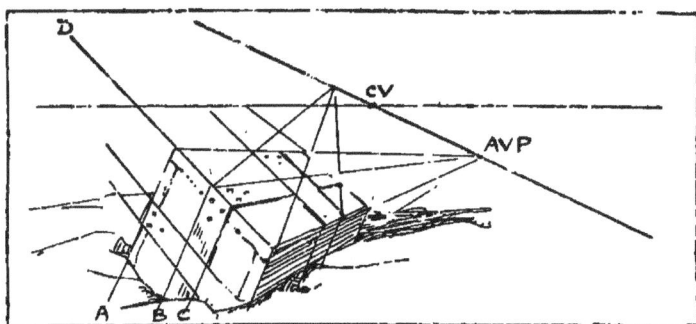

FIG. 28.—Lines in inclined planes perpendicular to the picture.

C are all perpendicular to the vanishing line. Line D,
representing the right angle to the lines going to A.V.P.,
vanishes somewhere out of the drawing on the inclined
vanishing line.

(*e*) *Directly ascending or descending planes.*

These planes resemble the ground tilted up or down.
The perpendiculars vanish on the line running from the

C.V.L (centre of vanishing line) through C.V. (centre of vision).

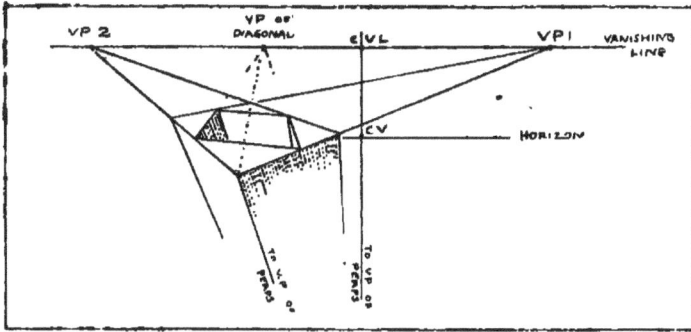

FIG. 29.—An object the top of which is in a directly ascending plane.

### (*f*) *Obliquely ascending or descending plane.*

Suppose we are sketching a chest floating in the water.

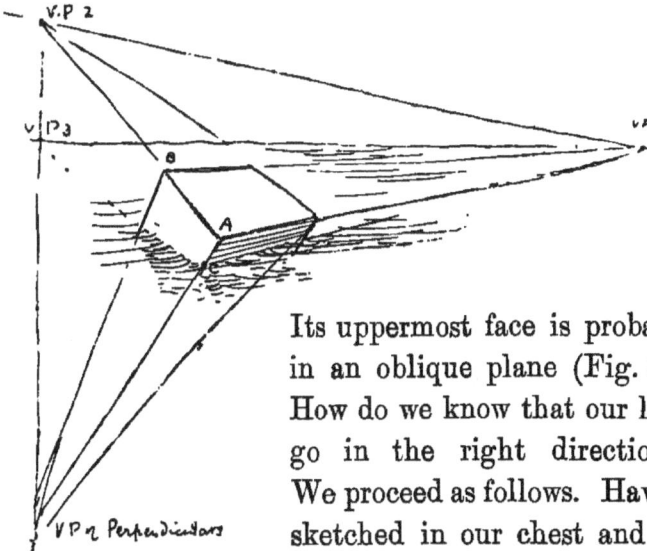

FIG. 30.—An object the top of which is in an obliquely ascending plane.

Its uppermost face is probably in an oblique plane (Fig. 30). How do we know that our lines go in the right directions? We proceed as follows. Having sketched in our chest and decided that the top is "about right," or looks pretty much

as we want it, we join up the sides. These meet at
V.P.$_1$ and V.P.$_2$, points of the existence of which we
had no previous notion. Obviously these are two points
in the vanishing line of the plane. The next question
is of great importance. Does either of these points lie
on the horizon? V.P.$_1$ does. This is accidental, perhaps, but the fact remains. If V.P.$_1$ be on the horizon,
then V.P.$_1$ is the V.P. of the direction the oblique plane
will make by its intersection with the ground-plane,
which in this case is the surface of the water. All planes
except horizontal planes intersect or pass through the
ground-plane somewhere; but it is only in the case of
the oblique plane that the whereabouts is important.
Why? Because the *inclination* of the oblique plane—
its tilt up from the ground—is at right angles to its
intersection with the ground. It is important, then, to
know which point represents the height of the plane.
If V.P.$_1$ represents the direction on the ground, its
reciprocal V.P., V.P.$_2$ will represent the altitude. This
will be clear enough when these points come to be found
later on by working.

Now, if V.P.$_2$ represent the height, it might be pushed
up and down to vary the inclination of the plane. So
that if V.P.$_2$ represents the inclination upward, the
inclination downward will be found below it, say at " V.P.
of perpendiculars " on the diagram. There is no way of
getting this by sketching, except the usual combination
of guess-work and " look right." The vertical line from
V.P.$_2$ cuts the horizon at V.P.$_3$. The plane of the side
of the chest cuts down through the water, and the water-

line will be that vanishing to V.P.$_3$. We prove this as follows. Assume that the line from A downwards to V.P. of perpendiculars cuts the water at C. Where will a line from B cut the water? We cannot as yet find out, because we have no line on the water to stop the line from B on its downward course. But V.P.$_2$ is a similar point to B—that is, it is in the same line. Now, a line from V.P.$_2$ to the VP of perpendiculars is precisely the same kind of line as we have from A and want at B. This line cuts the ground (for the horizon is on the ground even if it be at the farthest edge). We have, then, V.P.$_3$ and C both on the surface of the water. We join them, and they give the water-mark in the side of the chest. The water-mark on the other side goes to V.P.$_1$.

When considering Fig. 30, we came to a question of great importance. Was either of the vanishing points on the horizon? In that case one was. Now we have an instance where neither is on the horizon. We have, as before, our horizon representing the sea, with a cliff or two, and our chest as before. But joining up its sides, we find V.P.$_1$ comes below the horizon, V.P.$_2$ above. Neither is on the horizon. This will not matter as far as the top of the chest is concerned, but it will affect the perpendiculars. For we have just seen that the inclination of the plane was closely related to the trace of the plane on the ground, and consequently with a V.P. on the horizon.

All vanishing lines which cross touch where they cross. We need not, therefore, be in much doubt where the V.P. of the trace of our oblique plane on the ground will be.

It will be V.P.$_3$, where the vanishing line crosses the horizon.

We now guess V.P.$_4$, a reciprocal V.P. to V.P.$_3$; that is,

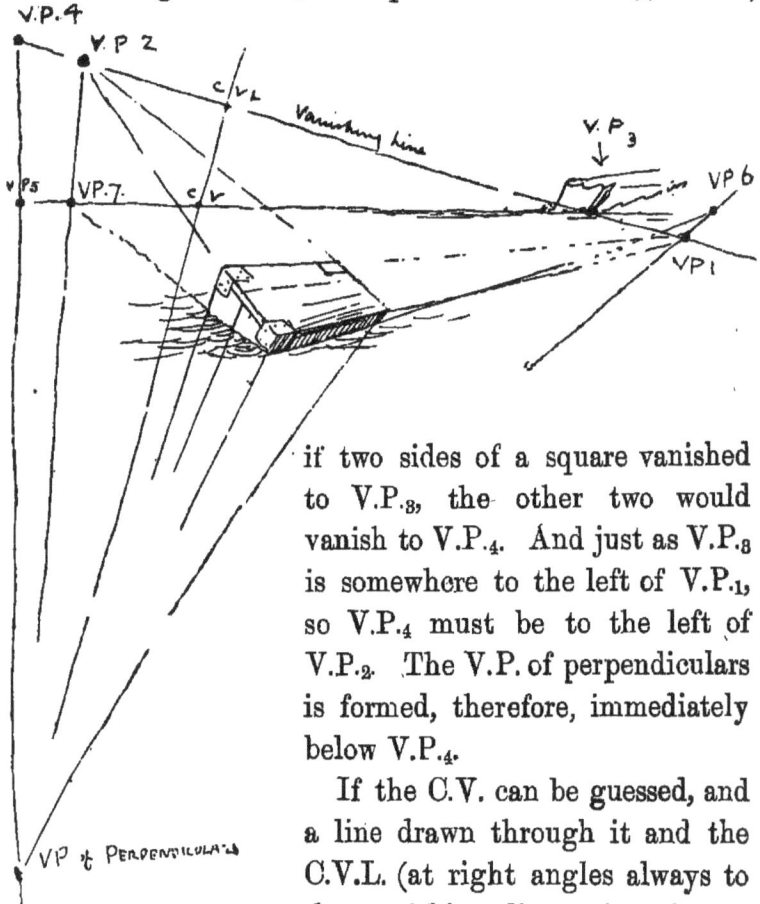

FIG. 31.—A chest floating in the water; its upper surface is an oblique plane.

if two sides of a square vanished to V.P.$_3$, the other two would vanish to V.P.$_4$. And just as V.P.$_3$ is somewhere to the left of V.P.$_1$, so V.P.$_4$ must be to the left of V.P.$_2$. The V.P. of perpendiculars is formed, therefore, immediately below V.P.$_4$.

If the C.V. can be guessed, and a line drawn through it and the C.V.L. (at right angles always to the vanishing line), then it too will point to the V.P. of perpendiculars. This is rather a matter for a later stage. C.V. is the centre of vision, and C.V.L. the centre of the vanishing line, and the point on that line nearest the spectator.

The water-line will also have to be re-considered. It will not be V.P.$_5$ (the V.P.$_3$ of Fig. 30), because the left side of the chest is not in the vertical plane through V.P.$_5$ and V.P.$_4$. It is in the plane through V.P.$_2$, and the water-mark will vanish to V.P.$_7$; on the other side it will vanish to V.P.$_6$, which is found by driving a line from V.P. of perpendiculars through V.P.$_1$ to the horizon.

### (g) *Planes parallel to the picture.*

All lines, no matter at what inclination they occur, which are in planes parallel to the picture are held not to vanish according to the theory of perspective.

True as this matter is, if we assume a picture-plane which is absolutely flat (and it were folly to work with any other), artists do not recognize this rule. They would never draw an entablature such as No. 1 of Fig. 32. They would admit No. 2, because it is directly opposite the spectator, and its lines may be expected to lie level; but they would continue to "dodge" No. 1, the left corner of which appears too high. They would droop the lines slightly to the left.

The diagrams in Fig. 32 are as follows: Nos. 1 and 2 are similar architectural details, both with their front surfaces and lines in the same planes parallel to the picture. The square ABEF is in a plane parallel to the picture; AB and EF are vertical lines, and do not vanish; AE and BF are horizontal lines, and do not vanish; all these are drawn with the T-square. The line EA is prolonged, still in the same plane, to I. IP is a slanting line in the same plane, for P is in the same plane as AB, and so is I. The line EA is also prolonged, the other way, to R. ER is the same length in fact and in the perspective drawing as AI ; there is no vanishing or diminution at all.

The side of the object ABCD is another rectangle, perpendicular to the first. It is in a vertical plane receding directly from us, and therefore goes to the centre of vision, C.V. A vertical line drawn through C.V. will give the vanishing line or far edge of the plane of ABCD, as of all other planes parallel to it. On this vanishing line will occur the vanishing points of any lines in the same plane as ABCD. AC and BD, as well as others, vanish there at the C.V. The extension of AC to K gives AK equal in fact to AI, but here shown smaller, because the line KAC is subject to the influence of vanishing. KP will also be in the vertical plane. It is shown continuing by a dotted line downwards, and will strike the vertical line through C.V. some inches below C.V. What is true of this plane in which ABCD lies is true of planes parallel to it. Thus EG and FH both vanish to the same point as AC and BD, that is, to the C.V. Similarly, LQ will vanish to the same point as KP.

A third plane is represented by the line MIJO. We have, of course, horizontal planes in the figure, the base FBHD and the top containing all the points R, L, E, G, N, K, A, M, I, J, O, and C. Of these N is the centre of the square EAGC, and by means of it we push out the diagonals to get M and O. If we join M to P and O to S, we gain the mitre lines where the two sloping surfaces LKQP and IJPS intersect.

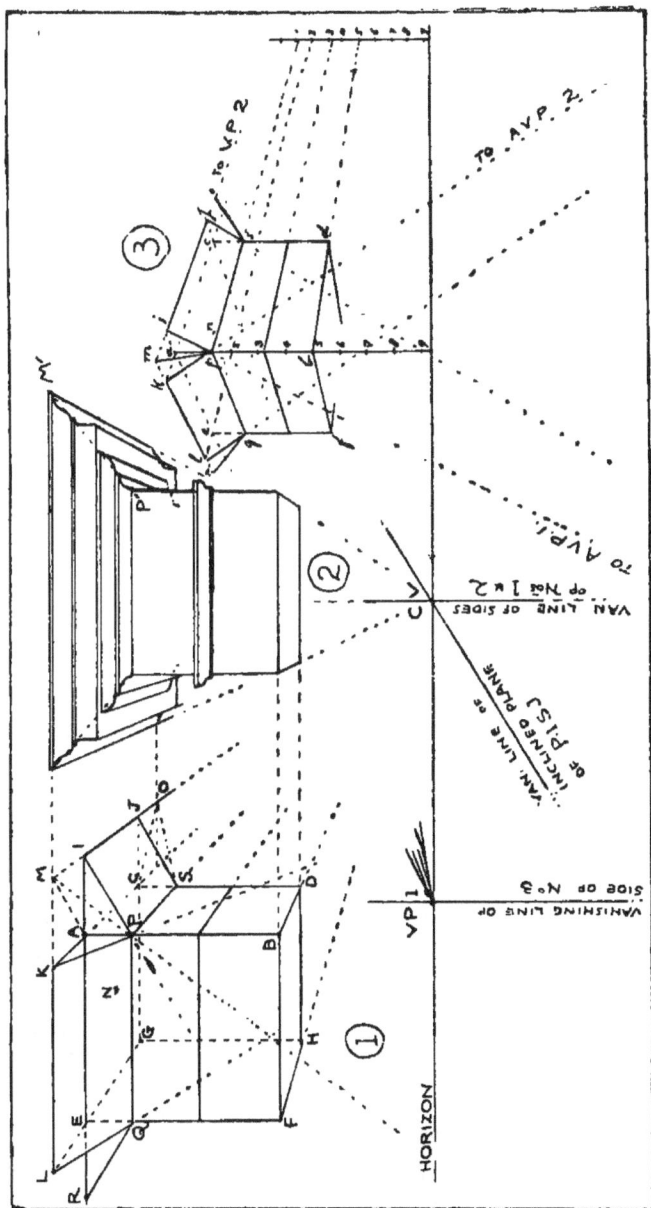

Fig. 32.—Objects partly parallel to the picture. Nos. 1 and 2 are in parallel, No. 3 in angular, perspective.

What kinds of planes are these sloping surfaces in? LKQP is in a directly descending. It is *directly* descending because its trace LK is parallel to the picture, and the perpendicular to this trace KP is running directly away just as KA is. If the lines KP and LQ were continued till they met somewhere below C.V., the point of convergence would be a point in the farther edge of the sloping plane. The farther edge (or vanishing line) would be parallel to the nearer edge LK. The angular line MP is also in this directly descending plane, and its V.P. would be on the vanishing line just described.

SO, the other angular line, must not be confused as going to the same vanishing point as MP. The plane in which SO occurs is similar in its relation to the object to the plane of MP, but instead of descending the plane of SO *ascends*. S is nearer to us than O, while M is nearer than P. M'P' in No. 2 would, however, go to the same vanishing point as MP in No. 1.

The plane in which PISJ, the sloping surface of the side occurs is an inclined plane perpendicular to the picture. Its vanishing line is drawn through the C.V. parallel to PI. PI has already been seen to be in a plane parallel to the picture, so that PI and SJ do not vanish or converge. The lines MP and SO are also in this inclined plane, as well as in the descending and ascending planes. MP will find a vanishing point in the V.L. of the inclined plane, somewhere on the left, below and outside the drawing. It can, of course, have only one vanishing point, and consequently it will be where the vanishing lines of the directly descending and the inclined planes cut one another.

In No. 3 the same kind of object, rather wider, is shown. It is, however, not in parallel perspective; only its vertical lines are in planes parallel to the picture as well as in other planes. Its front and side vanish to right and left. The V.P. on the left is found to be V.P.$_1$; that on the right is inaccessible on the paper.

A ready method of keeping lines which should converge in due difference of inclination when their V.P. is not available, is employed here. The slope of the top line being fixed, draw a vertical to the horizon. Divide this line by any ready scale, say $\frac{1}{2}$ or $\frac{1}{4}$ or $\frac{1}{8}$ of an inch. Find a similar vertical line which can be divided in the same way, but by a smaller scale. It is obviously an advantage to have the scale of the smaller line half the size of the scale of the longer. Of course, if zero vanishes through zero, and $9\frac{1}{4}$ through $9\frac{1}{2}$, then 3 will vanish through 3, or $4\frac{3}{4}$ through $4\frac{3}{4}$.

The working of No. 3 is the same as that of No. 1, except that lines which in No. 1 are parallel, and therefore drawn with a **T** or set square, are here vanished.

The planes resulting from this new position are as follows. Both front and side are in vertical planes at an angle to the picture. The vanishing line of the plane of the side is shown vertically through V.P.$_1$. On it *ip* will find its V.P. in accidental vanishing point 1. The plane in which *lkqp* occurs is no longer directly descending, but obliquely descending. Its oblique vanishing line would run through V.P.$_1$ and A.V.P.$_2$. Similarly the other plane, that of *ijps*, is an oblique descending plane instead of a laterally inclined plane as in No. 1. Its V.L. would run through V.P.$_2$ and A.V.P.$_1$. The sloping surfaces on the back of the object are in obliquely *ascending* planes.

One can decide, therefore, whether the roof of a house in one's drawing is in an oblique plane or in an inclined or directly ascending plane by noticing whether the cubical part of the house vanish to right and left, or is partly vanished to C.V. and partly parallel.

The student will have noticed that where we have a directly ascending or descending plane connected with the front of our cube (as in No. 1), we have an inclined plane perpendicular to the picture for the side. So that *inclined* planes answer to *direct* planes. In No. 3, as in all such cases, *oblique* at the side is answered by *oblique* at the front.

## Guessing Measurements.

The draughtsman who sketches in perspective has no guide but his eye and his common sense as to the dimensions of the objects he draws. *Freehand calculation* would be a better term, perhaps, than *guessing* for the process by which dimensions are obtained when nothing is to guide one.

The method rests entirely upon (1) the power of drawing regularly diminishing stripes, and (2) of seeing when a form supposed to be a square looks square. If a draughtsman cannot radiate contiguous stripes (as in Fig. 33), or see whether what he calls a square looks to be a square, he can do no sketching whatever, and probably no art work at all of any value.

In Fig. 33 it is evident that the lines of brickwork in C will transfer to the further end of that wall and pavement the sizes seen at the nearer. If the divisions between the bricks are known to occur every so often,

FIG. 33.—Guessing measurements.

these dimensions can be transferred from one end to the other by means of the converging lines. Similarly, the proportions of the house can be found by finding the relation between the dimensions of it and of the figure. Assuming that the man be 6 feet high, the 6 feet can be transferred to the corner of the house by assuming a "wall" vanishing from the man to the horizon, and taking the angle of the building in its course. The "wall" must be made to vanish on the horizon, because the man and the house are both

E

standing on a horizontal plane. A capable draughtsman would not need to draw the lines right to the horizon; but, of course, to take the lines to the horizon would make them more certainly right.

The six divisions representing the six feet should be guessed. With a little practice, the draughtsman will get to be able to guess dimensions with sufficient accuracy. A and B are two squares. The sides of A are divided by halves, the sides of B by thirds. Halves and thirds only need further like subdivision to yield fourths, sixths, eighths, ninths, etc. Fifths and sevenths must be guessed without previous subdivision.

## 18.

## Radial Projection.

To sketch perspectives one needs no great amount of knowledge, because in a sketch deficiencies do not seem to matter very much. If the drawing *looks* right, it *is* right. There are, however, many facts which are useful to the sketcher—these will become evident as the art of working a perspective is unfolded.

Whereas in sketching all has been guesswork based upon certain knowledge, and depending largely upon skilful drawing and an alert recognition of the relations of the various lines to one another, in working a perspective certainty is substituted for guessing, though the need for the understanding of the relation of lines one to another is not less, but rather greater.

Such perspective drawings as are made for use are more or less mixtures of working and sketching. The working of perspectives is rather laborious, and complicated subjects would take too much time to execute, were it not possible to substitute sketching for working when the details are reached.

A " worked " perspective is simply obtained by finding where the rays of light from the numerous corners of the object intersect the picture-plane.

Of course the result cannot be achieved without a complete knowledge of the geometric conditions. The process is, after all, only a kind of solid geometry done by radial instead of orthographic projection.

The simplest method is, therefore, that of pure radial projection. This is illustrated in Fig. 34. It will be seen that there is a plan of the object placed in its proper position toward the picture-plane, which is shown in plan. The spectator, or rather station point, is given also in due relation to the object and the picture-plane. Rays of light from all the corners are then drawn to the station point, and allowed to intersect the plan of the picture-plane. The convergence from right and left is thus secured. Dotted lines are then dropped down to carry the results of the convergence down the paper. The next step is to find the location, on these dotted upright lines, of the points they represent. This is done by another similar process. A side view or elevation of the object, picture-plane, and spectator is obtained. The same rays are drawn from the corners to the eye as before, and their intersection with the picture-plane noted. They are then carried horizontally across the paper till they cut the lines descending.

The system, although theoretically very sound and logical, is not at all useful. Not only is it laborious,

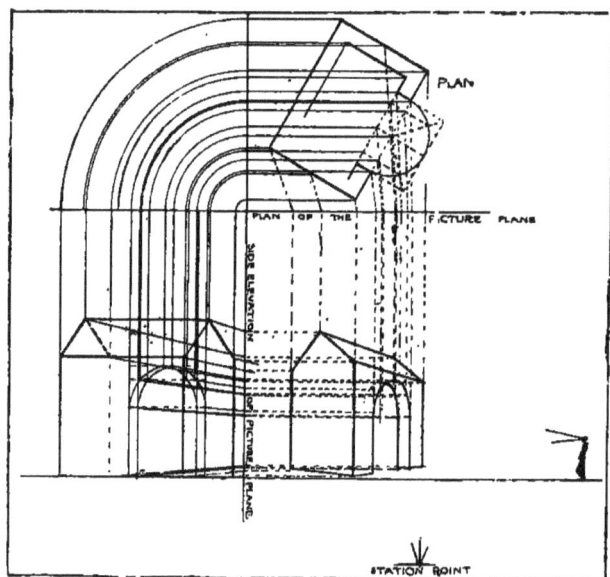

Fig. 34 —A perspective obtained by radial projection.

but it is treacherous. This defect is more to be noticed in the side elevation, when the rays come so close together as to be very troublesome, and even to render the process unworkable.

The plan part of the process is, however, not so dangerous, and is employed in architects' perspective, which is treated of in paragraph 24.

19.

## Vanishing Lines and Picture Lines.

The customary method of working perspectives differs from the radial, which was noticed in the last paragraph, in that, instead of getting the intersection on the picture-plane of rays conveying *points*, it gets the intersections of sheets of rays conveying planes. The customary method does not get a point, but a line, and if a point has to be found in that line it is obtained by crossing the line by another. Where the two lines cross is the point desired. The advantage of this method is that, since many points occur on the same line, with less labour one can obtain the general position of many points at the same operation. Another great advantage of the method is that it forces converging lines to converge properly. In the radial method a slip in working will throw the convergence out; in the customary method this is impossible, or highly improbable.

The process is based upon getting in the picture-plane the beginnings and the ends of lines of indefinite length. If we know where a line begins and where it ends, we can draw the line. If we know where a plane begins and where it ends, we can indicate the plane.

We have dealt on former pages with the seven positions which planes can occupy, and by a number of sketches have illustrated objects lying in those planes. All that was guesswork. The vanishing lines, or back edges, of

those planes were put where we thought it most likely
they would be, but there was no guarantee of correctness.
If the drawings look right, they are right. The same
must be said of the vanishing points. They are guessed.
Of course, where the form of the object is not pre-deter-
mined, where it is not obliged to be this or that shape
or size, our guessing and imagination can settle matters
between them. When, however, we are given such instructions as these—the object *must* be 15 feet in front
of the spectator, it *must* be 6 feet, it *must* be 3 feet below
the level of his eye, it *must* vanish at 40° to the right,—
then we have to work by more precise methods.

Lines are always conceived as being *in planes*. They
stretch from the front edge of the plane to the back edge
of the plane. This will probably make itself clear as we
proceed. We start always with the front and back edges
of planes. The front edges are called picture-lines, the
back edges vanishing lines. The method of obtaining
a vanishing line (V.L.) or a picture-line (P.L.) is not
always exactly easy, but it is also sometimes very far
from difficult.

20.

## The Horizon and the Ground-line.

Vanishing lines are the far distant or back edges of
planes. Hence we have a vanishing line, V.L., for the
plane of the ground. This V.L. is, however, called the

horizon, because it corresponds with the natural fact known by that name.

The front or near edge of all planes is where they are cut by the picture-plane. Usually we allow ourselves to think of planes beginning at the picture-plane and running back to their vanishing lines ; for the picture-plane gives us our basis of operations, and whether the ground-plane starts at the picture-plane, or at the

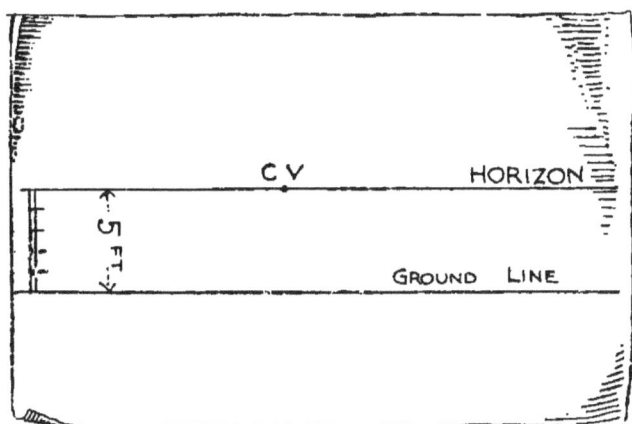

FIG. 35.—Front and back edges of the ground-plane.

spectator's feet (station point), or behind the spectator, or nowhere in particular, our operations *must* commence where the picture-plane cuts the ground-plane. Where this occurs is the ground-line (G.L.) or the picture-line.

Such, then, are the front and back edges of the ground-plane, or more correctly the picture-line and vanishing line, or again, more specially, the ground-line, and horizon of the ground-plane. Fig. 35 shows these lines as we should draw them when working a perspective drawing.

The horizon is drawn arbitrarily. It is our line of demarcation between "above" and "below" the eye— for it is on the level of the eye. We do not find it by any method, but take it as our first line, and draw it across the paper, putting C.V., for centre of vision, upon it, if we want to indicate exactly which point is immediately opposite the spectator. The C.V. indicates by fixing the middle which is the "right" and which is the "left" of the spectator (see Fig. 7).

As for our ground-line, we need to know how far by actual measurement the ground-plane is *below* the eye. Often the height of the spectator is given, say 5 feet, and this is held to indicate that the ground-plane is 5 feet below the eye. Of course the ground may be much further below the eye.

We obtain our ground-line by measuring down from the horizon the distance the ground-plane is below the eye. We mark down our 5 feet by some scale we determine to adopt, whether 1″ to the foot or ½″ to the foot, or whatever it may be.

It is reasonable to put down this actual measurement in this way, because we have to begin our practical operations *at* the picture-plane. Students sometimes think it would be more reasonable to take the actual measurements *at the spectator*. This only shows how little they realize that the picture-plane is our representative picture.

In Fig. 36 the height of the spectator is repeated by numerous vertical lines, all of which are 5 feet. If rays be drawn from the eye through *a, b, c, d,* etc., so as to cut the picture-plane, *only* the vertical line which happens to be in the picture will record on the picture the true height of 5 feet.

And while this is true of the vertical measurement downward from the level of the eye to the ground, it is true of *all* measurements. Measurements are actual *in, and only in, the picture-plane.* It does not matter in

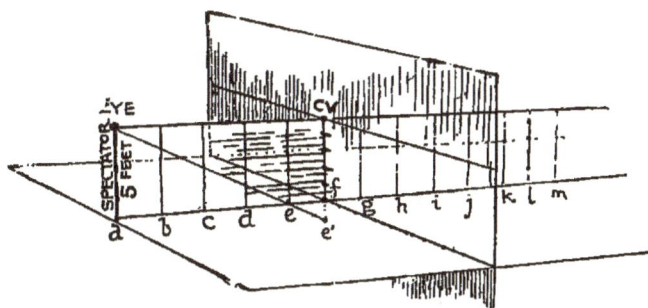

FIG. 36.—Measurements actual in the picture-plane ; all others reduced or enlarged by the rays radiating from the eye. The ray through *e* to *e'* alone is drawn.

what direction the line occurs, so long as it is in the picture-plane, and anywhere in the picture-plane, it is measured by the exact scale of feet or inches decided upon. Of course, whether the scale be $1''$ or $\frac{1}{2}''$ to the foot, or more or less, will make no difference except in the size of the drawing.

## 21.

## The Theory of the Horizon and of Vanishing Lines and Points generally.

We did not find the horizon by any method, but drew it across the paper anywhere likely to suit. If, however, we find it by the method we have to adopt for all other vanishing lines or back edges of planes, we proceed as follows. We prepare a side view of the whole situation,

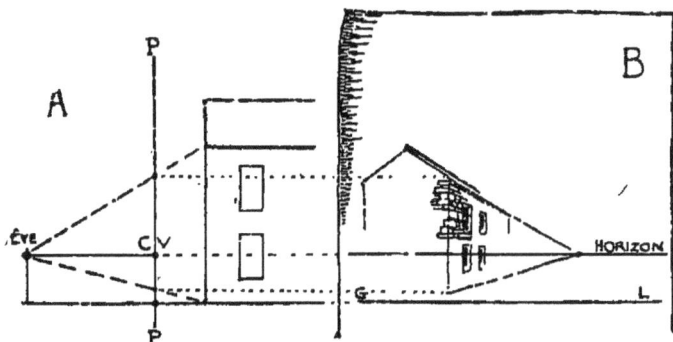

FIG. 37.—The theory of finding the horizon.

object, picture-plane, and eye, as in A, Fig. 37. We draw from the eye a line parallel to the plane whose V.L. we require. The plane is the ground-plane on which the house stands. We draw the line P.V.R. parallel to the ground-plane, and so obtain the point C.V. on the section of the picture-plane P.P. B represents the picture-plane seen in front view; the horizon is drawn across it level with C.V. on the other drawing. G.L. is also drawn level with the point where the ground-plane is seen to cut the picture-plane on the side view.

Accordingly the horizon occurs low down in the paper, because the eye is necessarily low down in relation to the house.

This process seems to achieve nothing. This is because the lines happen to run parallel with lines we already have, so that the horizon is at the level of the eye, and no fresh fact seems to be obtained. However, the method is precisely the same as that used in all cases, and usually gives new lines.

What has been done in the above case is this. *A vanishing parallel* was drawn parallel to the linear representation of the plane whose V.L. was required. The rule and process are the same whether it is the V.L. of a plane or the V.P. of a line which is required.

The theory is as follows. Let A and B in Fig. 38 be the edge view of two planes parallel to one another, or let them be two lines parallel to one another—it matters not which they are regarded as. To the spectator they will appear to meet.

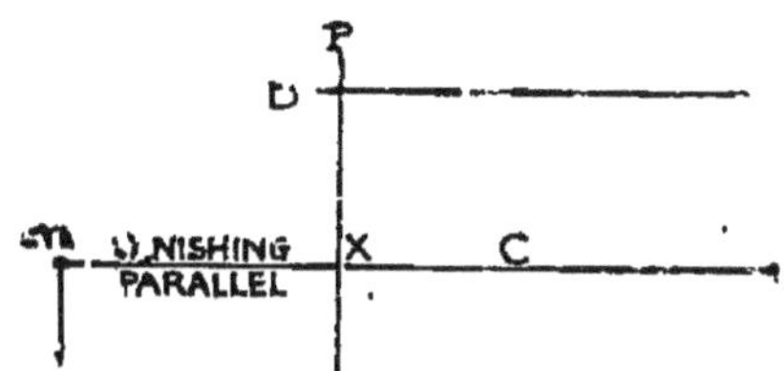

Fig. 38.—Finding the V.P. of A and B.

B will descend; A will ascend. Where will they meet?

It is a law of perspective *that parallel lines vanish in the same point*, and the law applies to any number of parallel lines. What we want to get is the record on the picture-plane of the distant point of convergence of the lines B and A. Now, if we assume a third line C, parallel to B and A, it also will converge to the same point. If we assume it immediately opposite to, or from, the eye, the law will still hold good, and, moreover, being placed to the eye like a gun to a marksman's eye, the whole line will be reduced to a point. The beginning and end of the line will, to the spectator, be one and the same. To the spectator, then, X, where the line C cuts the picture-plane, is as much C as any part of the line. It represents the extreme point of the line C, and is therefore its vanishing point. The same thing is done in Fig. 39,

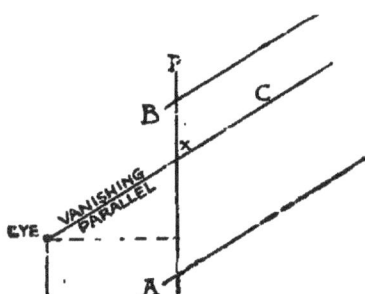

FIG. 39.—Finding the V.P. of A and B.

where A and B are not horizontal; X is their vanishing point. AX on the picture will hold all there is of the line A when carried through the picture to the eye. The student should be able to see that no straight line from the eye to the line A could ever get above point X, nor could a line carrying the most distant point of line B to the eye ever get *below* point X. So that BX represents the whole of B even if B be miles long. Of course the lines AX and BX are not throughout their courses of equal value. Near A the line AX does not represent much distance, but as X is approached very short distances indeed stand for very long ones on that part of A which is distant from the picture-plane. Thus we find the record on the picture-plane of the point of convergence of lines in perspective *by assuming a line parallel to them, and therefore repeating their conditions ;* a line, moreover, which, being assumed at the eye, is so placed that it appears to the spectator as a point only, and thus is the central line of that set of parallel lines of which it is one. Only one line in a set of parallel lines can appear as a point only, and if all the lines of the set converge to *a* point, that point must be that which this, the central line of the set, appears to be or is reduced to.

**22.**

## Finding Vanishing Points on the Horizon.

The last paragraph has been devoted to explaining how theoretically we find the horizon; how, in fact, we find the vanishing line of a plane. Possessed of our vanishing line, or horizon, if it be the ground or other horizontal plane we are dealing with, we have now to see how to find *vanishing points*, the terminations of lines lying in the plane. The process is precisely the same as that explained in the last paragraph. *We assume at the eye a line parallel to the line we wish to find the V.P. of, and produce it till it cut the horizon.*

Fig. 40 will perhaps sufficiently illustrate how the matter stands. We see there the eye out before the picture-plane as usual. Straight before the cye is the centre of vision, C.V. The plane carrying the angles cuts through the picture-plane, and makes a level trace, horizon, upon it. It is clear that lines could be set off at any angle, and that they would strike somewhere upon

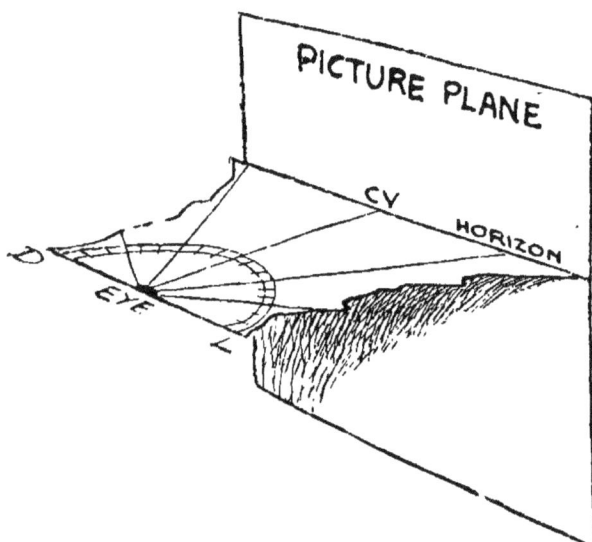

FIG. 40.—Angles set off from the eye.

the horizon. The diagram shows where 60° to the right and 60° to the left would strike it.

But there is a technical difficulty to get over. The eye is out in front of the paper of the picture-plane. We have, therefore, to drop it down into the plane of the picture as shown in Fig. 41.

Through the eye will be noticed the directing line D.L., against which the angles are set. It is parallel to the picture, and appears in Fig. 42, where we have the working lines as we use them.

We are able, therefore, now to obtain vanishing points

on the far edge of the ground.  The process as it works
out in practice is shown in Fig. 43.

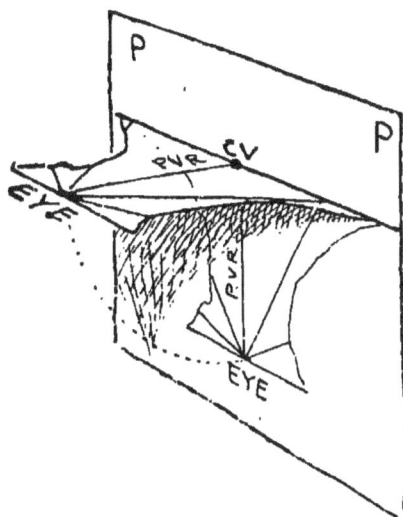

Fig. 41.—The eye and the angles
laid down into the picture.

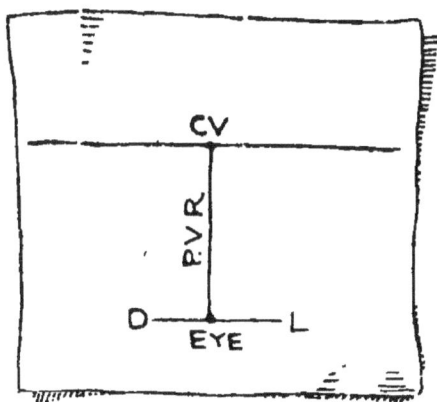

Fig. 42.—The working lines.

There we have the eye set down as far below (it might

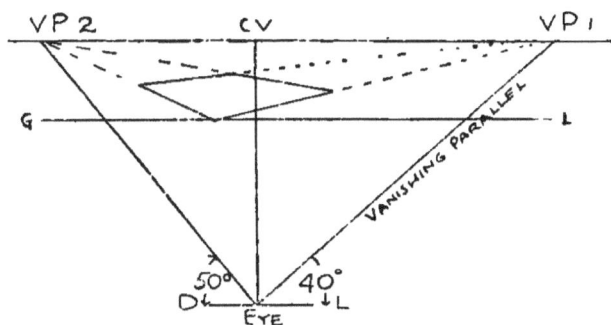

Fig 43.—Finding vanishing points.

as well be above) the C.V. as it properly is before it.
We have the D.L. parallel to the horizon.   Against

this directing line are set up angles 40° on the right
and 50° on the left, which we will suppose are the angles
required.    These vanishing parallels, we note, are at
right angles to one another, and will thus give the
vanishing points to right and left of a square object.
The vanishing points are shown, and a rectangle in the
ground-plane vanishing to them.

23.

## The Line of Height.

Vertical lines do not recede, or are held not to recede.
Measurements upon the same vertical line are therefore
equal.    It is one of the defects of the science of perspective that lines parallel to the picture do not suffer
any change as they stretch farther and farther upward,
or to this side or that.    The diminution is, however, so
slight naturally, that our rules would become so complex
as to be impracticable were any attempt made to gain
this diminution.

With all its defects, therefore, we have our line of
height, or *line of elevation,* as the "Jesuit" called it.
Height is the stated altitude of an object when measured
directly upward from the ground.    Hence a height is
always represented by a vertical line.

But if a vertical line itself does not suffer any diminution in its dimensions by reason of one part being higher

up than another, lines of the same height, but at different distances within the picture, appear at different lengths. If we imagine a row of such lines placed so that the row recedes into the picture, then we may draw through the upper and lower ends of the verticals lines representing, one, a trace on the ground, and the other, a distance of so many feet above it. In a word, we draw a wall, with the same height throughout, though gradually diminishing in measurement, on the paper. The top and foot of this wall, being parallel horizontal lines, will vanish somewhere on the horizon.

The upright, where the wall begins, is the line of height, the wall we may call the *wall of height*. Upon the wall, like lines of masonry, can run any dimensions we like.

In the Jesuit's "Perspective" the wall of height is placed at the side quite independent of the object drawn. The reason for this was chiefly to do away with as many working lines in or on the drawing as possible. The author of that work vanished his wall to any point on the horizon.

He has a diagram to which Fig. 44 is similar. That the method very greatly facilitates the getting the several heights in such an instance as this, there can be no doubt whatever. In his diagram, as in Fig. 44, the height does not begin at the nearest upright object, but at one halfway round on the left.

In another case of a winding stair, the Jesuit has this line of elevation, or rather wall of elevation, at the left side at the foot of the square "well" within which the

F

stair winds. The steps are situated at the diagonals and diameters of the square. The wall of elevation is for one stair only, but as it recedes it passes all the different locations backward in the picture, though on the ground. When the steps some distance from the ground have to

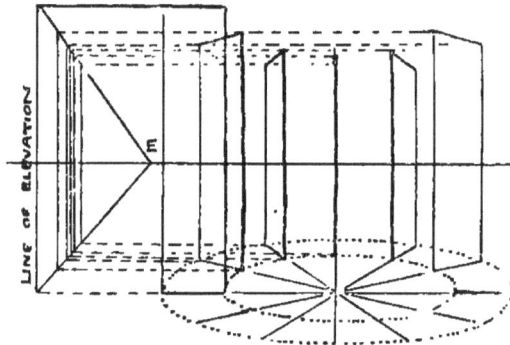

Fig. 44.—Several vertical rectangles standing in a circular arrangement with their heights obtained from one wall of height.

be measured, he returns with his compasses, and picks off the size he wants according to his need at the moment,

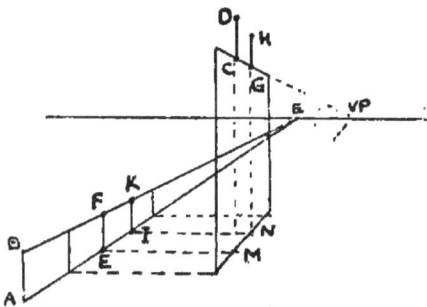

Fig. 45.—Measurements transferred from a "wall" to higher positions at the same distance.

from whichever part of his wall is at the same distance back as the step he is busy with.

The gist of his method is given in Fig. 45. The height of CD is required to be the same as AB. It is therefore taken by the compasses from EF. The points E, F, M, C, and D are all at the same distance within the picture EM, being parallel to the horizon.

In modern practice it is not usual to bring the wall of height down from any point, such as E in Figs. 44 and 45, as did the Jesuit, but to bring them down from one of the V.P.'s. The modern practice avoids the many horizontal lines across from the object to the wall of height; but if it avoid some lines in this way, it uses the slanting lines of wall in the middle of the working with some little danger of confusion. The student should only follow the Jesuit's practice when such figures as Fig. 44 are being worked, and I expect an examiner would even then prefer the modern method.

Nevertheless the "wall" will be more readily understood from the reference made to the Jesuit's method, and the method itself may sometimes prove useful.

## 24.

## Architects' Perspective.

The method employed by architects for making their perspective drawings is given in Fig. 46. The several drawings there shown in one diagram usually exist on three or four different sheets. The two elevations and the plan would be on separate sheets, and be part of the set of drawings prepared for the erection of the house.

Across the finished drawing, and stretching out beyond

it, will be seen the horizon, its V.P.'s being connected with eye 2 below. A little below the horizon is the ground-line. These are the machinery by which the diminution inwards, or vanishing, is obtained. At point C a wall of height is set up containing the vertical dimensions of one wall of the house. These dimensions are taken from the elevations which are here given below. Such walls of height are obtained as required. The distances above the ground and their inward diminution being provided for, the distances apart from one vertical line to another have to be obtained.

This is done by means of the plan and rays radiating from it to the eye. Eye 1, as it is marked on the diagram, is placed the proper distance before the plan of the picture-plane AB, which is itself placed immediately against the plan of the house. The plan of the picture-plane could be placed at any distance before the eye, even behind the plan of the house. But wherever it be situated, the distance of the eye before it must be repeated when on the other drawing the V.P.'s are found. In a word, eye 1 must be distant from AB, the same distance as eye 2 is distant from horizon.

The angle at which the plan of the house is placed against AB depends upon the choice of the draughtsman, who determines which view of the building will best suit his requirements. The position of the plan against AB corresponds with the angles drawn from eye 2, or rather, the angles from eye 2 are drawn parallel to the sides of the house on the plan.

It will be readily seen how the method works out.

Rays are drawn from points on the plan to eye I, and cut the plan of the picture-plane AB. Thus diminished

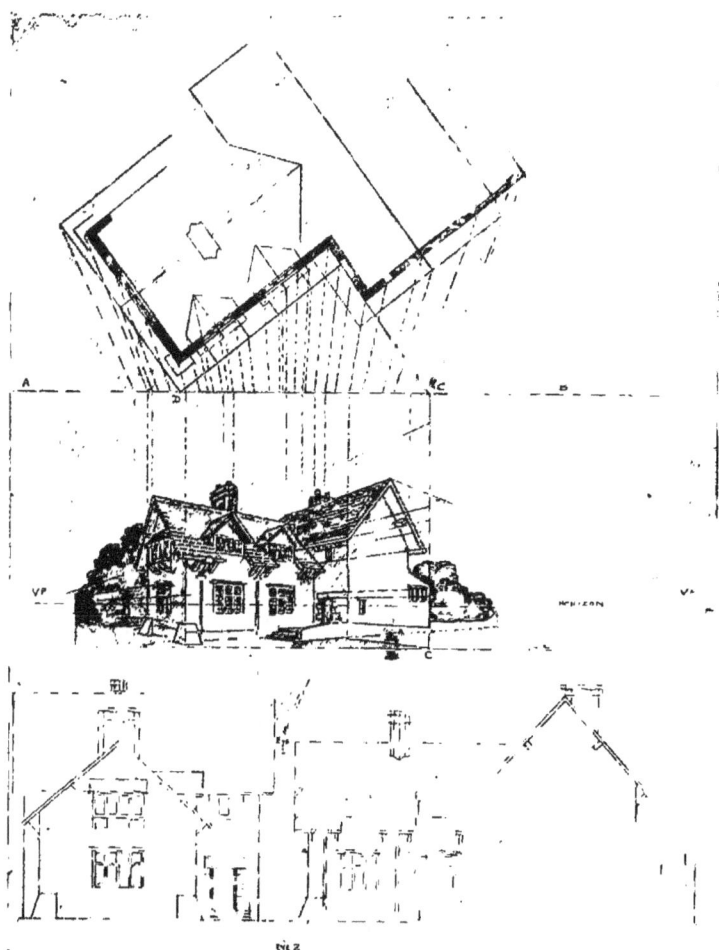

FIG. 46.—Architects' perspective.

laterally, the dimensions are dropped vertically down across the perspective drawing till they cut the lines vanishing to one of the V.P.'s on the horizon

The method is a mixture of the methods used in Figs. 34 and 43.

Architects also draw their perspectives by the method in which measuring points are used, and which is the method chiefly advocated in this book. A few further observations on the practical application of that method will be found in the paragraph on perspective for designers, etc. p. 136

25.

## Measuring Points.

We have learned in paragraph 21 how to find vanishing points. We are able, therefore, to draw a line from its commencement at the ground-line, where it starts, say 4 feet by actual measurement to the left of our centre-line, to its ultimate end in its vanishing point. We have now to find out how to measure distances upon that line by some other method than that of radial projection. Now, if we can get one line we can get another, and we can so arrange this second line that it crosses the first at a point we wish to locate. This is done in Fig. 47. At the side of that diagram is a plan of the picture-plane, and touching it a plan of a line AB, which makes the angle Y with the picture-plane. The centre of the picture-plane is indicated by C.V., and A is seen to be distant to the left the distance M.

If we wish to get a perspective representation of this line AB, we repeat these conditions on our perspective drawing. We set off the distance M from the centre line, along the ground-line, towards the left. We next set off the angle Y at the directing line, and so obtain the far end of our line, namely its V.P. We then draw the line. Now, if we want to measure the distance AB along A, we return to our plan and assume any line BC, and obtain the perspective representation of it just as we

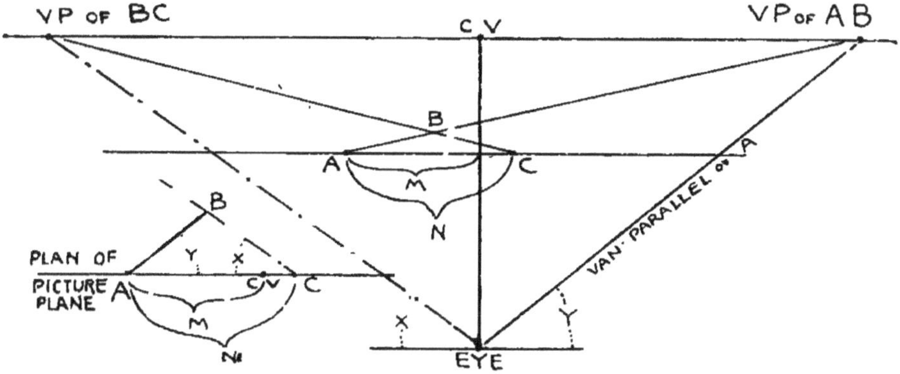

FIG. 47.—One line crossing another, and so indicating a measurement upon it.

did that of AB. That is to say, we set off the distance N from A on the ground-line toward the right. This gives us point C. We then draw a vanishing parallel for BC, that is, we set up the angle X from the directing line. We thus gain the V.P. of BC, and so can draw the line. Where the lines from A and C cross is the point B, the distance required.

This process involves, however, two actions which can be avoided: (1) we had to measure how far C was from A;

and (2) we had to measure the angle X. With lines taken by chance, this must always happen.

If, however, the assumed line be so taken that it forms the base of an isosceles triangle with AB and AC for legs, all difficulty is avoided. This is done in Fig. 48, where the assumed line BC is the base of the isosceles triangle ABC. There is no need to measure AC, because, as a rule, we know the length of AB in feet and inches, and AC is equal to AB. It will be seen that the vanishing parallels for AB and BC repeat, with the horizon, the

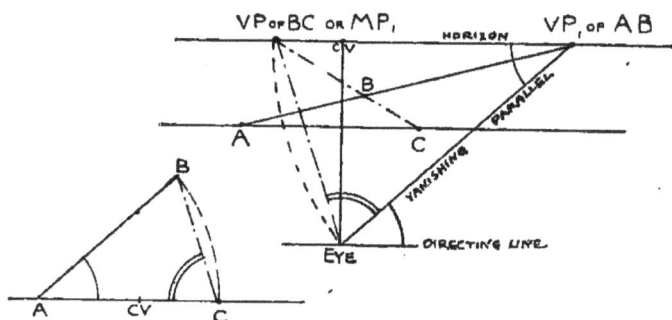

FIG. 48.—The theory of the measuring point.

conditions of the isosceles triangle on plan. From the measuring point to the vanishing point is the same distance as from the eye to the vanishing point. The vanishing parallel runs from the eye to the V.P.; it runs from a line through the eye, the directing line, to a line through V.P., the horizon, and these two lines (directing line and horizon) are parallel. The alternate angles on either side of this vanishing parallel are therefore equal; this is indicated by the single curves spanning the angles. Now, the angle at the eye was made equal to

angle BAC on plan, hence the angle M.P., V.P., eye, is
equal to the angle BAC on plan. The base BC, on the
plan, was obtained by striking an arc from A as centre
and AB as radius. In the same way, on the working
drawing, we place the point of the compass on V.P., stretch
the pencil to eye, and describe the arc, which finds for us
the point M:P. on the horizon. The chain line from the
eye to M.P. is thus the vanishing parallel of the chain line
BC on plan. M.P. is, therefore, the V.P. of an assumed

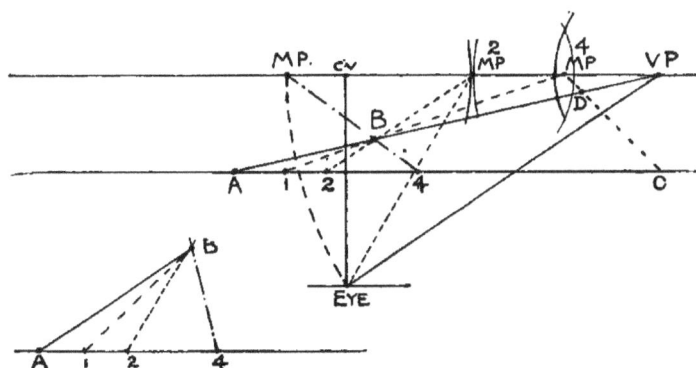

Fig. 49.—Measuring long distances.

line, BC crossing one line A, and crossing it in such a
manner that it records on the picture-plane AC, the same
length as it serves to measure on AB. It is a point, too,
found by the ready means of describing an arc from the
eye to the horizon, using V.P. as a centre.

It will readily be seen that we can invent measuring
points, to some extent, to suit our purposes. Thus we
can find points which measure twice or four times
the distances set along the picture-line. This is shown
in Fig. 49. We have seen, by what has been said,

that the distance on the horizon from M.P. to its V.P. corresponds to one of the legs of the isosceles triangle on plan. Thus in Fig. 49 it corresponds to the leg A4. If we divide A4 into halves, and so obtain point 2, we get a new assumed line, 2B, which measures AB twice A2, because AB and A4 are originally equal. We obtain the vanishing point for 2B by merely dividing the length from M.P. to V.P. on the horizon in half. We thus obtain a measuring point which measures *double*. By repeating the process we get another, M.P.4, which measures fourfold. By using this measuring point, we obtain AD

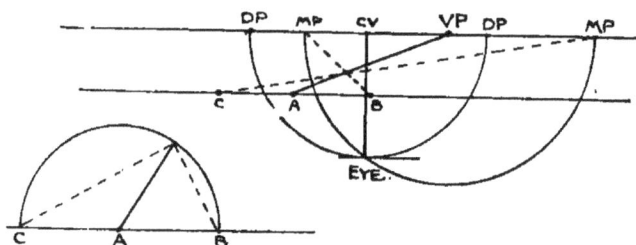

Fɪɢ. 50.—Two M.P.'s to each V.P.

*four times* A.C. The usefulness of these additional measuring points (*fractional*, they are sometimes called) is very evident, when we consider that it would have been almost impossible to measure AD by the ordinary M.P.

*Every V.P. has its own M.P.* is a rule never to be forgotten. A measuring point will only measure lines, but of course *any* lines, converging to its vanishing point. We customarily speak of *the* measuring point, as though there were but one for each vanishing point. There may be two, as Fig. 50 shows. There A, in plan, is the line, B is the assumed line measuring it. It is the

base of an isosceles triangle, and its equivalent M.P. is
on the left of the C.V. in the worked drawing. But C
is also the base of an isosceles, giving CA equal in length
to the line, and is produced, like B, by an arc struck
from A. In the worked drawing its equivalent M.P. is
out on the right, usually too far off to be of any use.
When the C.V. is a vanishing point, two measuring
points are, however, usually found for it. These are

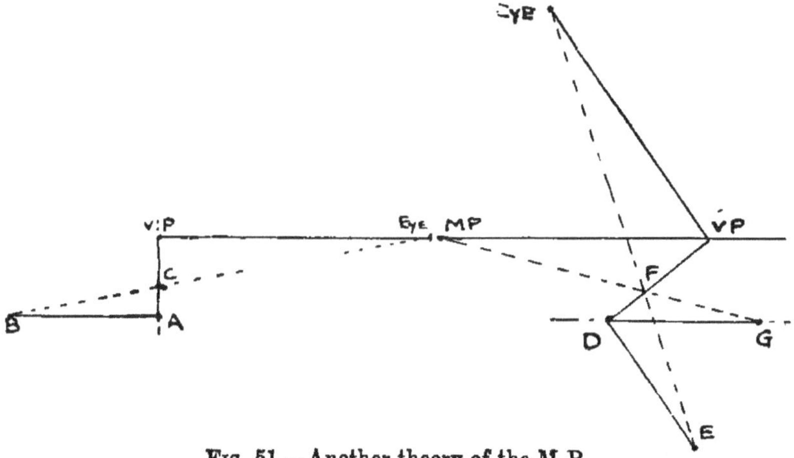

Fig. 51.—Another theory of the M P.

commonly called *distance points*, because they measure
the lines which gives distances within the picture, distance being indicated by lines perpendicular to the
picture-plane, and therefore vanishing to the C.V.

Another theory of the measuring point can be given.
It is illustrated in Fig. 51. There, on the left, is a
section of picture-plane, eye, and object. Suppose AB to
be immediately in front of the eye, so that the vertical
from A to V.P. can act as a swivel. The ray from B to

eye will give C, so that AC is the perspective length of AB. If this occurred in the middle of a drawing with V.P. at the C.V., then we can readily imagine the lines turned round as on a swivel, so that eye fell upon the horizon, and B came on to the ground-line. How would this act if the line A to V.P. were not at right angles to AB? On the right side of the diagram we have D, and a line from it to V.P. If we put the length of the vanishing parallel out from V.P. to eye, and the length of the line out perpendicularly, DE, we can get our ray from eye to E, and measure F. And the same result is obtained if the lines V.P. to eye and D to E are at an

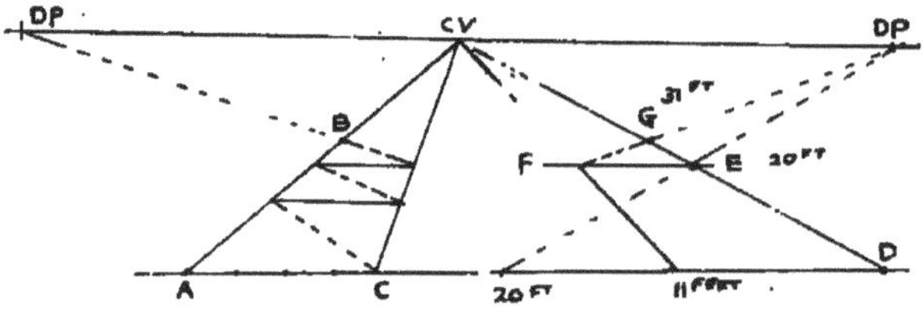

FIG. 52.—Long distances found.

angle to DF, so long as they themselves are parallel. This is shown by points M.P. and G.

Another method of finding long distances within the picture is shown in Fig. 52, and is taken from the Jesuit's "Perspective."

Line AB vanishes to C.V. DP is the measuring point. AC is 4 feet. AB is required 12 feet. From C a line is drawn to C.V., and a measuring line from C to DP. A point is thus obtained which is perspectively 4 feet up AB. At this point a line parallel to the ground-line is drawn to the line from C to C.V. This serves as a new ground-line, and is, of course, 4 feet long. The process is repeated till B is reached at 12 feet distance.

Again, suppose a distance 31 feet be required from point D. 31 feet could be set off from D toward the left, and taken to DP, but if this is to be avoided, the result can be obtained as in the diagram. A distance, say 20 feet, is taken and measured along D to E. At E a new ground-line is drawn. Along this new ground-line the distance 11 feet is found as shown. At F, and from F, the 11 feet are carried to G by the line to DP.

It is not necessary that the V.P. should be the C.V., or, in other words, that the subject should be in parallel perspective—but the Jesuit knew no other.

## 26.

### Parallel Perspective.

One rather wishes the term *parallel perspective* had never been introduced. The division of the problems of perspective into parallel and angular does not lead to the understanding of the subject it is supposed to facilitate. There is no difference between parallel and angular perspective either in theory or practice; the only difference is that the objects in parallel perspective are in positions which give more lines which do not vanish, but are drawn parallel to one another by the

T square or set square, than is the case with objects in angular perspective.

The rule by which we find vanishing points and vanishing lines is that *we draw from the eye a line parallel to the original line, till it touches the picture-plane.*

Now, some original lines are parallel to the picture-plane, and are in planes parallel to the picture; so that lines drawn from the eye parallel to them will never touch the picture-plane. Hence no vanishing point can be found, and the lines which thus have no vanishing points are as a consequence drawn parallel to one another in the perspective drawing.

The chief lines which thus do not vanish are vertical lines. They can always be assumed to be in planes parallel to the picture. But lines slanting to left or right may equally be in planes parallel to the picture. Such lines also do not vanish, but are drawn parallel to one another.

The fact has been several times impressed that perspective prefers to deal with cubical forms. What, then, becomes of a cubical form when it is as much as possible parallel to the picture? If one side is parallel, where does the other go? Straight forward. If we then apply our rule for finding the V.P. of the sides which go straight forward, we find they go to the C.V. Hence in parallel perspective the C.V. looms large.

The facts of the situation, so far as plan is concerned, are shown in Fig. 53. Here ABCD is the base of the cube; AB and CD are parallel to the picture, AB being

in it. They neither of them vanish; any attempt to
draw a line from the eye to the picture-plane (which is
shown in plan) parallel to CD must fail. But the vanishing point of AC and BD can be found. A line from the
eye parallel to them brings one to the C.V.

The measuring point for these lines vanishing to the
C.V. will yield the diagonal to the square. AD is the
diagonal. AD is the base of the isosceles triangle, by
which BD would be measured, BD and BA being the

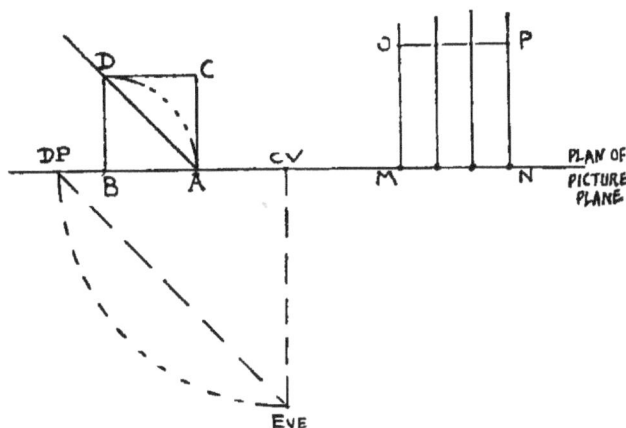

Fig. 53.—The theory of parallel perspective.

equal legs. The vanishing point of this diagonal—the
measuring point, that is—of all lines going to the C.V.
is found, theoretically, by the line drawn from the eye
parallel to AD, and yielding D.P. (or distance point),
and found practically by striking an arc with C.V. as
centre and the distance to the eye as radius.

It will be guessed why this measuring point is called
a distance point. Distances *within* or beyond the picture
must necessarily be represented by such lines as AC,

lines running straight forward. As these lines are
measured by the distance point, it is obvious that the
chief use of that point will be to secure distances before
the spectator.

It must be noted, also, that the four lines on the right
side of the diagram enclose the same distances between
O and P as between M and N. Were the lines not
perpendicular to the picture-plane, this would not be

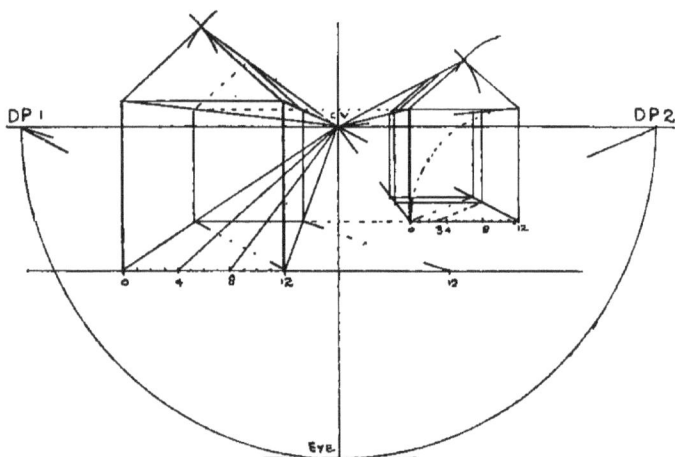

Fig. 54.—Objects in parallel perspective.

the case; then the distances between the lines would
be less than those found where the lines impinged upon
the picture at M and N.

In Fig. 54 is given a simple example of parallel perspective. On the left is a cube with 12-feet sides. The
12 feet are marked off along the ground-line. The
distance back is found by the diagonal drawn to the
distance point, DP. The sloping top or roof has edges
of 8 feet. These distances of 8 feet are put in by arcs

of circles of that radius. The figure on the right is done in the same way. The 12 feet are transferred from the ground-line; but once the front of the object is reached, its front line can be divided into feet, and serve as a scale from which the upper part is worked. This is done, and a distance of 9 feet is cut off. Compare the street scene, Fig. 105.

The picture-plane need not be *before* the object. We are therefore free to work forward out before the picture-plane, if it facilitates our work.

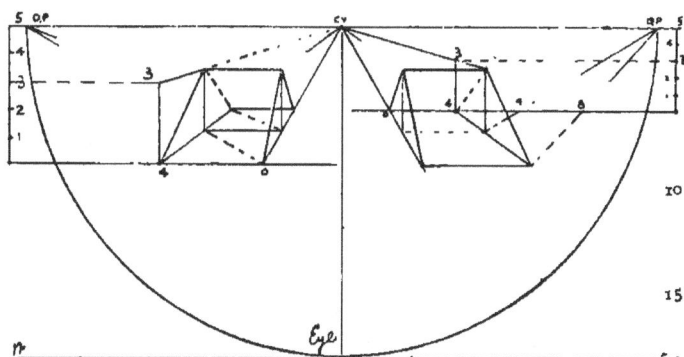

FIG. 55.—The picture-plane before and behind the object.

This is shown in Fig. 55. There we have an object with base 4 feet by 8 feet, and 3 feet in height. The ground-plane is 5 feet below the eye, and the spectator is 12 feet before the picture. The ground-line is drawn, therefore, 5 feet down by actual scale—the scale is shown at the side. On the right side of the diagram we have the same object in the same position (except that it is on the right of the spectator). Its nearest point is still 12 feet in front of him, but the picture-plane is shifted to the far end of the object, or 8 feet further away. The ground-line is drawn at the level of the far end of our first object, and the space it gives up to the horizon is divided at the side into a new scale of 5 feet. The problem is worked *forward*. The measurements are taken from the diminished scale. Without any change in our diagrammatic position of the eye, it is now 20 feet in front of the picture-plane instead of 12.

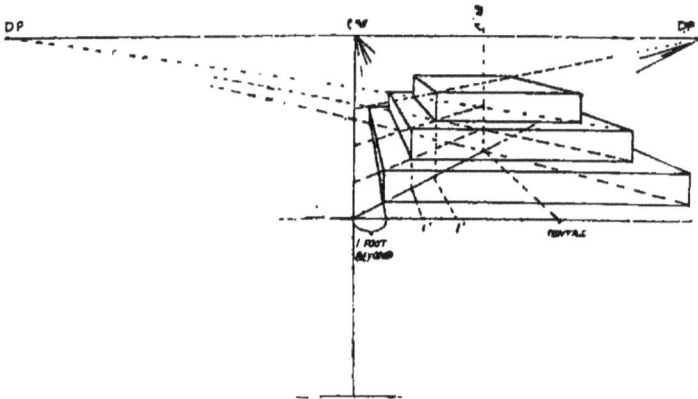

FIG. 56.—Three steps, the upper successively 1 foot within the area of the lower, and each 1 foot in height. The front edges are parallel to the picture. The distance points, which also give diagonals of the squares, are utilized for shrinking the upper steps within the lower. The diagonals carry the heights and positions out from the axis to the corners. The lowest step is 1 foot beyond the picture-plane. The nearest corner is 1 foot on the right.

Fig. 56 gives another example of parallel perspective, wherein the two distance points acting at one moment as measuring points for lines vanishing to C.V., and serving also as vanishing points of diagonals to the squares, very greatly facilitate the receding of the steps.

## 27.

## Parallel Perspective applied to Objects in Angular Positions.

Lines perpendicular to the picture and parallel to the picture, which are the proper subject of parallel perspective, are lines in the simplest conditions possible.

Lines in angular positions are not so simple, therefore, nor is their perspective representation so easily obtained. But the difficulties of angular perspective can be avoided (though the gain is a very doubtful one) by treating the object in parallel perspective. Fig 57 shows the representation of a hexagonal prism cut off above by a sloping section. The prism is seen in plan at the lower part of the diagram; it is standing on a horizontal plane, and is at an angle to the picture. A, B, C, D, E, and F are the six corners of the hexagon. To work the problem by parallel perspective, we neglect the fact that A, B, C, etc., form a hexagon; we merely note that A, B, etc., are at certain distances from the picture-plane, and we regard them as the terminations of lines, short or long, perpendicular to the picture-plane and at certain distances apart.

This diagram, apparently so elaborate, owes its complexity to its being also an example of a particular method of working, a method which was elaborately expounded by the Jesuit, Andrea Pozzo. The method is based upon the following principles: (1) that the

diminution in width will be the same if the object be
worked on a lower plane than it properly stands on, and
(2) that the diminution in height will be the same on a

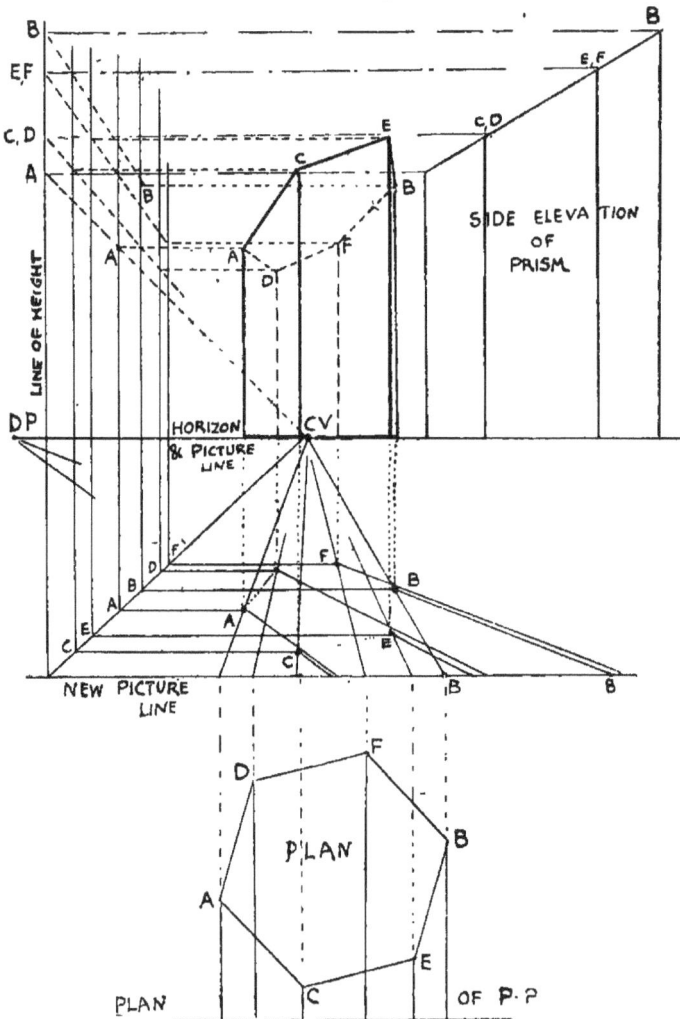

Fig. 57.—A method of Andrea Pozzo's.

line of height at some distance to the side of the object,
as in the proper position.

One incontestable advantage of this method is that by it an object standing on a plane at the level of the eye can be represented. In the present case the object is standing on such a plane, a plane of which the ground-line and horizon coincide. Pozzo points out that the method not only makes such a case workable, but that workings very much squeezed up are avoided. In Fig. 94 the lines cross so obtusely as to render mistake easy. To get out of the difficulty Pozzo assumes a new ground-plane as low down as he likes. On this he gets his figure in perspective as clear as in a bird's-eye view. The student must note that A, B, C, etc., are merely found as points in lines all vanishing to C.V. The distance point is used to measure them with. The working of B is shown throughout.

With the figure thus obtained on a low ground-plane the corners are thrown up into their proper position. Next follows the getting of the heights. The points A, B, etc., are thrown out sideways and then turned up an angle, and so make above lines of height. The heights are transferred from the geometric elevation on the right to the first line of height, and from it are carried by lines to C.V. to their proper lines. Then they are finally carried across to the figure itself.

Owing to the D.P. being placed so near C.V. to allow it to appear in the diagram, the distance of the spectator is too short to give a reasonable view of the object. This explains the curious aspect of the perspective drawing.

## 28.

## Angular Vanishing Points; Angular Perspective.

If by *parallel perspective* we mean the perspective of objects partly parallel to the picture, then angular perspective is the perspective of objects partly placed at angles to the picture.

The working of a simple problem of angular perspective is given in Fig. 58, and may be described as follows.

The horizon having been drawn, and the C.V. placed upon it anywhere, the eye is found by dropping a perpendicular from C.V., and making this perpendicular 10 feet long. Then the ground-line is drawn parallel to the horizon 5 feet by scale below it.

We next have to find the point at which the object starts. This is point A, 1 foot on the left. We therefore place point A 1 foot by measurement from the scale to the left of the central line. The object is not recessed within the picture, that is, its nearest point touches the picture, so that there is no pushing of point A back toward the C.V. to be done, as there would be if A were stated to be 2 or 3 feet "within" the picture, or "behind" the picture.

Point A being thus settled, we think of our cubical form, which nearly always makes itself felt. We find by the plan that the side AB vanishes at 40° to the right, and that side AC vanishes at 50° to the left. Our

vanishing points to right and to left have then to be
found at the angles named.

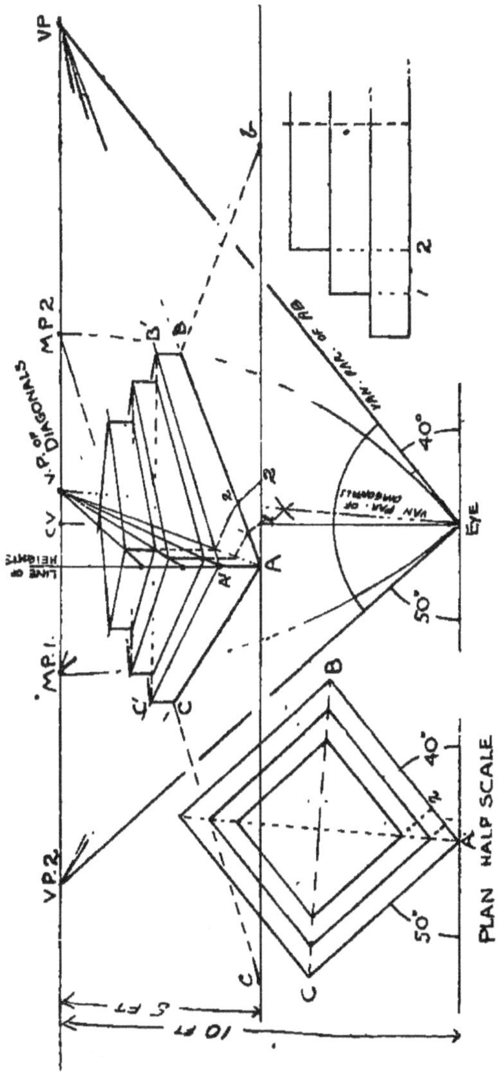

FIG. 58.—Plan (at half scale) and half elevation given of three steps, the upper smaller than the lower. The height of spectator above the ground is 5 feet, the distance of the eye before the picture-plane is 10 feet. The plan shows the angle at which the steps are placed to the picture-plane. Point A is 1 foot on the left.

We find the vanishing points by drawing "vanishing

parallels" at 40° and 50° at the eye. These give us V.P.$_1$ and V.P.$_2$.

We at once now find the measuring points for both of these. We do so by placing the point of the compass on each V.P., and describing arcs from the eye to the horizon. We notice that each V.P. has its M.P. on the other side of the C.V. to that on which it itself occurs.

We also obtain the V.P. of the diagonal. This is an auxiliary of great service. We obtain this by bisecting the angle between the two vanishing parallels.

We proceed to work the problem by drawing lines from A to V.P.$_1$ and V.P.$_2$. Then we measure along these lines the length of the lowest step. That is, we mark off along the picture-line A*b* a measurement taken from the elevation of our steps. From *b* we draw a line to M.P.$_1$, and this cuts the line from A to V.P.$_1$ in B, and fixes that point. C is found next in the same way. Having thus points C, A, and B, we could complete the square base, if we wished, by drawing from C to V.P.$_1$, and B to V.P.$_2$. These two lines would cross and give us the fourth corner of the square. In the working of this figure, however, all invisible parts which are not needed for the working are omitted.

Our next task is to raise the edges of the lowest step; to get, that is, the short vertical lines CC', AA,' and BB'. These vertical lines complete our cubic conditions. Being purely vertical, they do not vanish. We therefore erect short verticals on C, A, and B. A happens to be in the picture, so we can mark on the vertical over A the height of the step as it actually is, so that AA' is really the height of the step. From A' we draw to V.P.$_1$ and to V.P.$_2$, and so obtain the upper edges of the lowest step on the two sides near us.

We now have to fall back to the second step. This distance is shown in plan and elevation as 1 foot. We measure the distance A1 along AB in our perspective drawing. We do this by first marking A1 along the ground-line, and drawing from 1 to M.P.$_1$. This transfers point 1 on the ground-line to its position on AB, so that the two dimensions are equal, the one being the perspective equivalent of the other.

To complete the figure, we, in this case, make use of our diagonals. We have the V.P. of those diagonals which pass more or less directly away from us. The diagonal CB will vanish some long way out of the picture on the left. Since, however, we already have C and B, we can draw the diagonal CB without having recourse to its V.P.

Having thus point 1 (and point 2 is managed in just the same way and at the same time) on the edge of the square, we next transfer it to the diagonal from A by a line drawn to *V.P.*$_2$. Above the points on the diagonal we raise perpendiculars, which give us the near corners of the upper steps by intersecting with the lines of their height from the measured "line of heights" to the V.P. of diagonals. Carrying our lines right and left from this central "mitre," we limit them at the corners on right and left by the diagonal C'B' and the similar diagonal on the step above.

## 29.

## Examples of Angular Perspective.

The problem in Fig. 60 involves certain difficulties to which special reference must be made. The large square being found in perspective, the vertical stripes are found upon the lower edge by carrying the dimensions placed along the ground-line to M.P.$_1$, which measures the lines going to V.P.$_1$. In placing these dimensions along the ground-line, a commencement has to be made by *bringing a line down from M.P.*$_1$ through A of the square to A on the ground-line.

The horizontal stripes are carried by a kind of wall vanishing to V.P.$_2$ from the line of heights A6.

The student must carefully guard against two errors, if no more—

1. When, by means of M.P.$_1$, we carry the dimensions ABCDCBA to the first line of the large square, we draw certain dotted lines. *These will not measure the back square as well as the front one.* They will not at the same time give A'B'C', etc., for the reason that ABC on the ground-line are located there according to the position of A; and thus B'C', etc., must be located on the ground-line in accordance with A' if they are to be measured by M.P.$_1$.

2. Lines of height, such as A6, can be found by lines from any point on the vanishing line (horizon),

but care must be taken that they are not used
for obtaining heights in other positions than those

Fig. 59.—A circular diso with a square hole through it is given in plan
and elevation. Circles can only be found in perspective by taking
a number of points in the circumference and getting the perspective
representation of those. Through the points the curve is then drawn
by freehand. This method applies to any curve, whether part of a
circle or not. It will be seen that the square is *striped* vertically
and horizontally in such a manner as to give all the points, both of
the circle and the inner square which are needed.

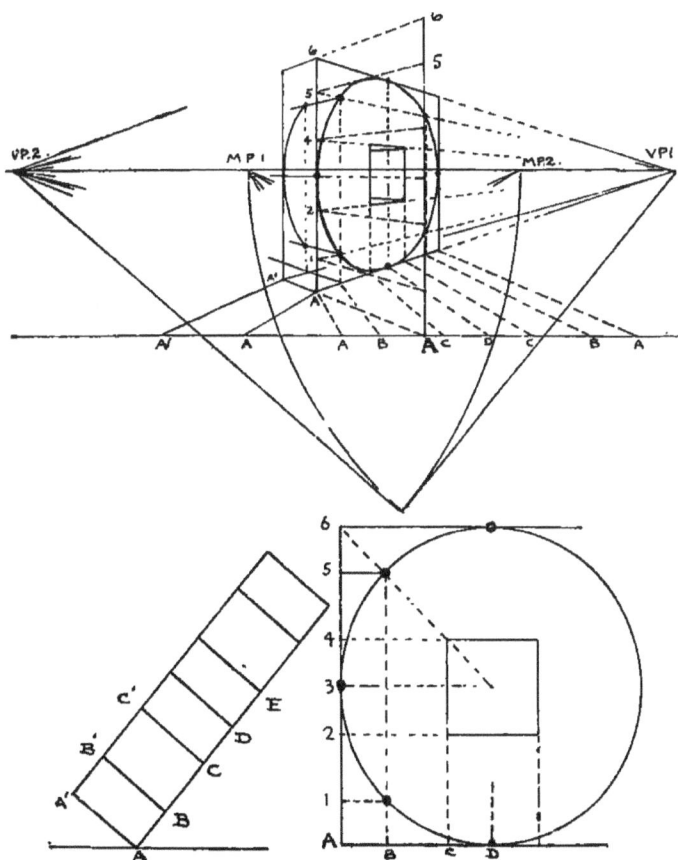

Fig 60.—Plan and elevation of the objects in Fig. 59.

of the points through which they are drawn. Thus our line of height is here found by bringing

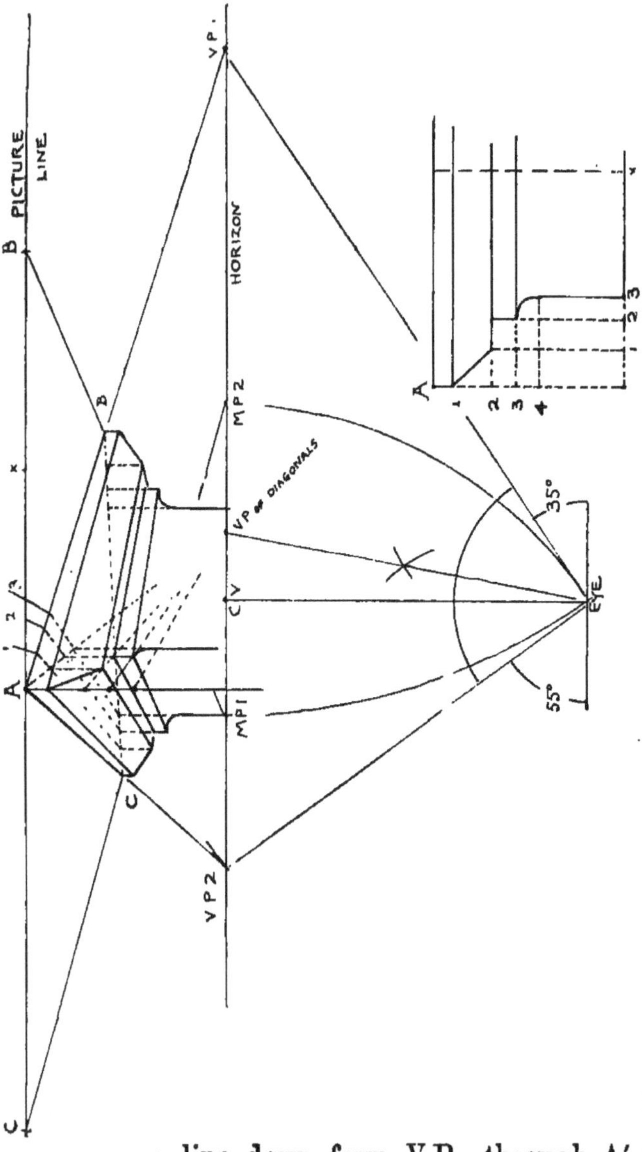

Fig. 61.—A square capital in angular perspective. The top of the capital is in a plane some distance above the eye, therefore we work from a *picture-line* instead of the ground-line. The side AB vanishes at 35° to right. The use of the diagonals is clearly shown. The points 1, 2, 3 are first transferred from the picture-line, where they have their actual size, to the line AB by means of the M.P. of V.P.₁, *i.e.* M.P.₁. They are then carried across to the two diagonals by lines parallel to AC, vanishing, that is, to V.P.₂. A line of heights dropped down from A gives the vertical measurement of the three points. These measurements are carried to the V.P. of the diagonals in order that lines may be dropped from 1, 2, and 3 on the diagonal above to touch them.

a line down from V.P.₂ through A′ and A to A

on the ground-line. Here the line of height is set up with its dimensions upon it. From the

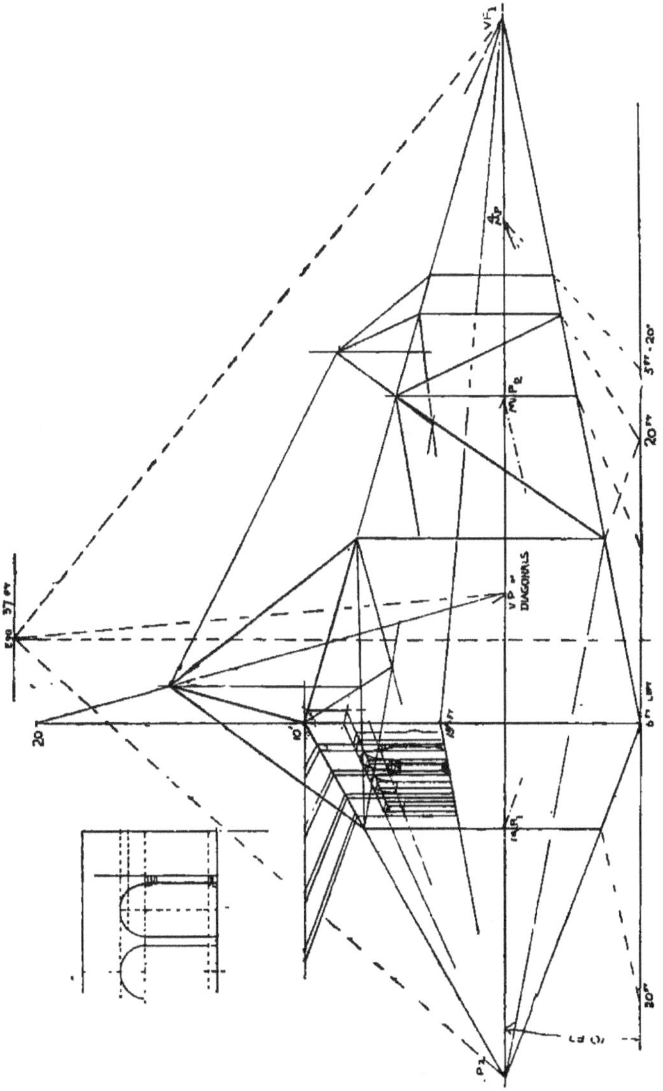

Fig. 62.—The upper part of a church. The distance of the spectator before the picture is 37 feet. The total height of each tower is 45 feet. Each tower is 20 feet wide, and the space between them 60 feet. The fractional M.P., $\frac{4}{M.P.}$, is used.

various dimensions lines are then returned to V.P.₂, the point from which the original line was brought down. The lines thus drawn are over one another, and of course have no power whatever to give height over other points.

## 30.

## Inclined Perspective.

What has gone before under the heads of *parallel* and *angular* perspective has to do only with lines parallel to the ground, horizontal as they are called. True, vertical lines were freely drawn, and lines of height used But these vertical lines did not vanish, and required no theoretical knowledge.

If the reader will turn back to the paragraphs on *sketched* perspective, he will find a tabulated series of planes lying, not horizontally only, but in seven kinds of positions.

We have now to see how we obtain the vanishing line and vanishing points proper to these planes with exactness instead of by guesswork.

The business is, after all, not one of great difficulty; indeed, it is almost as easy to understand the working of problems in this, that is sometimes called " advanced " and " third-grade " perspective, with its vanishing points up in the air and down in the ground, as it is the ordinary perspective, wherein only vanishing points on the horizon are used. There are difficulties of course, but they are really few, and should be methodically attacked.

The chief difficulties may be enumerated as—

(1) Finding the new position of the eye.

(2) Finding the picture-lines of the inclined planes.

The former of these need give no trouble whatever; the latter sometimes involves some little ingenuity, and helps to " muddle " the worker, but the writer's experience is that students as readily pick up the treatment of all kinds of planes as they do the horizontal ones merely.

## 31.

## To find a Vanishing Line.

The theory of all worked perspective is this. Let the perspective drawing be conceived as drawn upon a plane called the picture-plane. Those of the lines of the drawing which are supposed to be parallel in reality will have to converge to vanishing points, or they will not appear foreshortened. Lines on the same levels (in the same planes) will have their vanishing points in rows, which thus suggest the far edges of the planes. These far edges are called the *vanishing lines* of the planes. So, just as lines vanish to points, do planes vanish to lines. The theory of the working is this: If planes or lines parallel to the supposed real positions of the lines represented in the drawing on the picture be assumed at the spectator's eye, and allowed to touch the picture, they will touch the picture at the points or lines where the

lines or planes of the drawing will have to vanish. Figs,
38 and 39, pages 59 and 60, illustrate the application
of the principle. The theory is dealt with also there.

Accepting this rule and principle, what do we do when
we apply it?

We find we have a *picture-plane* and an *eye* somewhere
before it.

The least we can know, and the first thing we know is,
how far the eye is before the picture-plane. We know,
too, that the point in the picture-plane immediately
opposite the eye is called the C.V., centre of vision, and
that the line joining the eye to the C.V. is called the

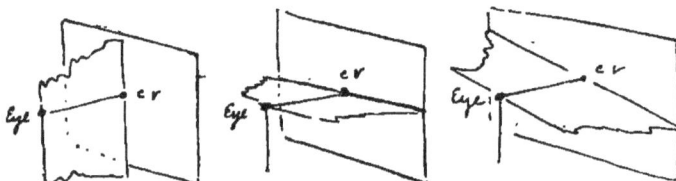

Fig. 63.—The vertical, horizontal, and inclined planes, all perpendicular
to the picture.

P.V.R., or principal visual ray. This P.V.R. is perpendicular to the picture-plane, it sticks straight out from
the C.V., and it neither bends upward or downward, nor
to right or to left.

It is obvious that any plane made at the eye directly
toward the picture-plane must include this P.V.R., and
must include the C.V. If we imagine planes passing from
the eye to the picture and including the P.V.R. and
C.V., we shall find they occupy *three* positions to which
names can be given. The vertical position is when the
plane is straight up; the horizontal, when level; the

inclined, when it is anywhere between the other two. Throughout this work an "inclined plane" must be understood as a *laterally inclined plane perpendicular to the picture.* All planes neither level nor vertical are inclined, but the word used by itself has the significance here defined. Of inclined planes fulfilling these conditions there may be any number, that is, the angle of inclination may vary, but there can be only one vertical plane and one horizontal actually through the eye and C.V. Where these imaginary planes from the eye touch the picture-plane are the vanishing lines of all planes fulfilling the same conditions as these imaginary planes.

How do we get these vanishing lines? Obviously, for the vertical and horizontal we merely draw down or across through the C.V., precisely as is done in the sketched examples in Figs. 25 and 27. Geometry in these cases is not called into requisition; we know the line goes through C.V., and that it goes vertically down or horizontally across, and the result is obtained at once.

In other cases the vanishing line has to be found by geometry. How, is explained in the succeeding paragraphs.

We have there the rules for finding the vanishing lines of the seven kinds of planes (see Fig. 22). There is really not very much to master. We see that this is so if we divide the seven into sets.

Three of the planes are perpendicular to the picture, and have their V.L.'s through the C.V. These three are shown in Fig. 64. They are found very simply by our drawing them through C.V., the horizontal one first

H

then the vertical at 90° to it, and then the inclined at *whatever* angle to the horizontal or vertical it is known to be.

We have thus dealt with—
 1. The horizontal.
 2. The vertical, perpendicular to picture.
 3. The inclined,  „      „      „
If now we mention—
 4. The plane parallel to the picture,
we find it has no vanishing points at all, and so is negligible.

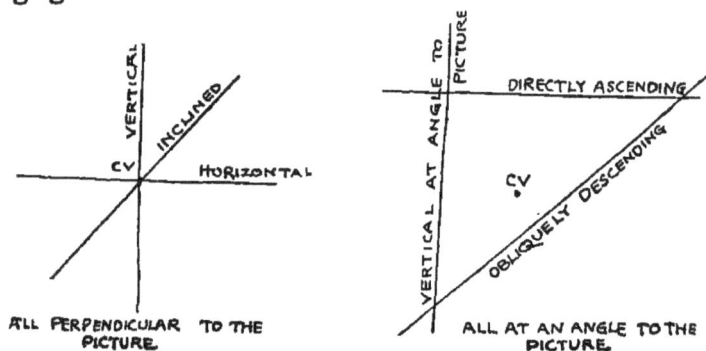

FIG. 64.—The six vanishing lines.

We have thus only three to seriously trouble us—
 5. The vertical at an angle.
 6. The directly ascending or descending.
 7. The obliquely ascending or descending.
We see by Fig. 64 that the vanishing lines of these are removed from the C.V.

## 32.

## To find the V.L: of Horizontal Planes.

Merely to maintain a systematic method which is valuable, we give a separate paragraph to this subject.

*Mark any point C.V. on the paper, and draw a line across the paper through it, using the T square.*

## 33.

## To find the V.L. of Vertical Planes, Perpendicular to the Picture.

These also have been sufficiently treated. The rule is—

*Draw through C.V. a line down the paper at right angles to the horizon if that be previously found.*

## 34.

## To find the V.L. of Inclined Planes, Perpendicular to the Picture.

These planes are usually described as at such and such an angle to the ground. When that is the case, the inclination is of course to the horizontal, and the angle is set up through the C.V. upon the horizon.

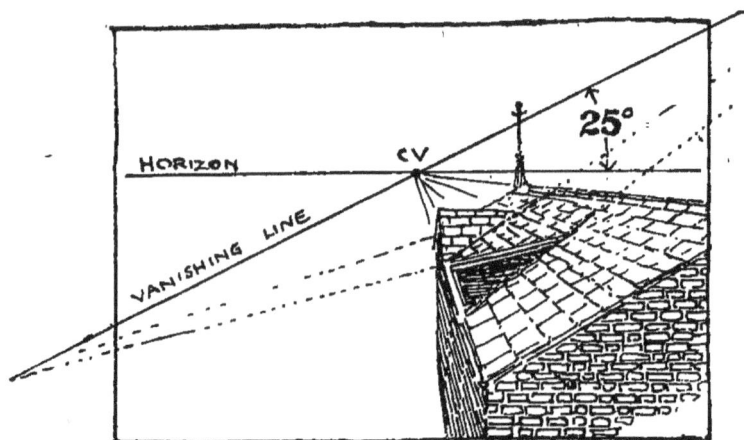

FIG. 65.—Vanishing line of inclined planes making 25° with the ground, and being perpendicular to the picture.

*Draw through the C.V. a line at the same angle to the horizon as the plane is declared to make with the horizontal or ground-plane.*

## 35.

## Planes Parallel to the Picture.

These have no vanishing points or lines, and the lines upon them are drawn without diminution. Of course the lines which are perpendicular to these planes recede and vanish. Hence some lines in Fig. 66 vanish. The arches are, however, done with rule and compass.

FIG. 66.—Planes parallel to the picture.

## 36.

## To find the V.L. of Vertical Planes inclined to the Picture.

To obtain the vanishing lines of this and the two following planes, we have to master one of the difficulties which trouble the beginner.

We can best clear away the mists by, in the first place, recalling the fact that vanishing lines are the remote or distant edges of their planes. We draw their *traces* on the picture-plane, not themselves. That is to say, we draw on the picture-plane lines which hide them and stand for them.

So that if, as in Fig. 67, we have a vanishing line some distance to the left of the C.V., we cannot say it is so many feet or inches or miles on the left. Both C.V. and this line are properly in the remote distance, and the *distance* between them cannot be measured. But if the V.L. cannot thus be found by setting off a

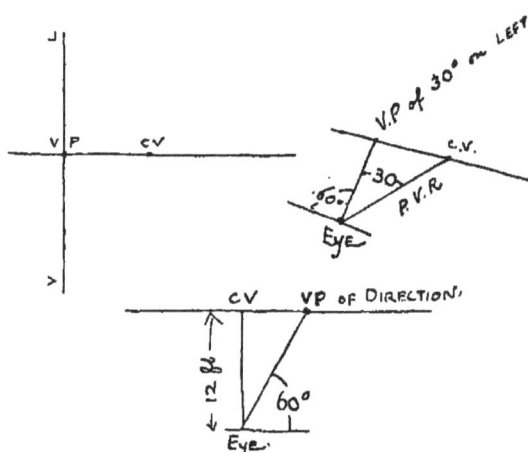

FIG. 67.—The theory of finding the V.L. of a vertical plane at an angle to the picture.

*measurement* from C.V., it can be found by an *angular* method.

We know that C.V. is directly in front of the spectator. Can we not know also in what similar relation the V.L. on the left is to him? If the P.V.R. (principal visual ray) represents the direction straight ahead, then a line to this V.L. on the left from the spectator will represent the *direction* toward the left. Hence we speak of planes "at 60° to the picture-plane toward left," and so on.

The angle used in description, it must be noted, is *not set down from the P.V.R.* always, but in some cases, and especially in the one now being considered, it is set out from the directing line running through the spectator's shoulders and parallel to the picture-plane. The student must be careful to watch that he does not fall into mistake in this matter. It requires constant watching, for the circumstances vary with the problems. Mistakes such as these come always from the inattention which one is easily guilty of in working perspective drawings.

What we have to do, then, is to find out the solution to this problem—

*If C.V. represents "straight ahead," when the eye is, say 12 feet before the picture, what point on the horizon will represent 60° on the right?*

We require a geometrical representation of the position. This is also given in Fig. 67. At the lower part of that diagram we have the C.V. and the eye separated by their 12 feet. We see also an angle of 60° set up from the directing line. The line (called a vanishing parallel) which terminates these 60° yields V.P. on the picture-plane. The lines stop at C.V. and V.P. on the picture-plane simply because in perspective, as has been so often said, we are frankly getting on the picture-plane the points and lines which stand for those in the far distance.

V.P. is then the point on the picture related to C.V. at 30° (60° at the directing line being equal to 30° to the P.V.R.) when the eye is 12 feet before the picture-plane. This angular measurement of 30° from the P.V.R. is also shown in another diagram in Fig. 67.

*V.P. being then a point representing* 60° *on the right, if
we require a V.L. for vertical planes at* 60° *to the right, we
have simply to draw a line vertically down through it.*

The student should hardly need telling that the
relation between a vertical plane and the picture-plane
is to be *seen in plan.*

## 37.

## Finding the V.L. of directly Ascending or Descending Planes.

The illustration, Fig. 68, shows a directly *ascending*
plane. Here the angular relation between the plane and
the P.V.R., which is the indicator of the level or hori-

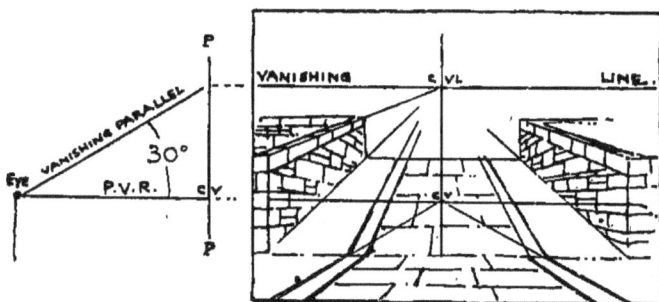

FIG. 68.—V.L. found of a directly ascending plane which makes 30°
with the ground-plane.

zontal position, is shown at the side of the diagram.
Note that the ascending or descending planes are described as at such and such angles to the horizontal

or ground planes. This, at all events, is the usual practice.

The method by which the result is obtained is so exactly that explained in the last paragraph in connection with the vertical plane, that there is no need to go over the ground again.

## 38.

## Finding the V.L. of obliquely Ascending or Descending Planes.

It will be necessary to commence by drawing attention to the characteristics of these planes. If an ascending plane, instead of proceeding straight forward in a direct manner, rises somewhat toward the left or right, it is called an *oblique* plane.

We see it in a perspective view in Fig. 69. The plane is shown something like a desk, making 30° with the ground-plane, and its front edge, or rather intersection with the ground, making 40° with the picture.

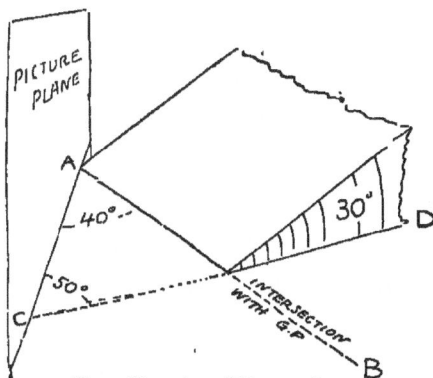

FIG. 69.—An oblique plane.

It is evident—

(1) That the intersection AB will *always* be on the ground, and therefore it may be followed right away to the farthest edge of the ground-plane, that is to the horizon.

(2) That the inclination (30°) must occur, and must be measured, on a line CD *perpendicular to AB.*

(3) It is evident, that while the farther side of the plane touches the picture at A, the near side will touch the picture some distance below C. So that when the plane strike the picture-plane, it will do so as an *oblique line* higher at one side than the other.

The method is shown in Fig. 70. The plane would be described as one *having its intersection with the ground, vanishing at* 40° *toward the right, and ascending at an inclination of* 30° *toward the left.*

Briefly, the method is a mixture of two actions. The V.P. of the direction of the intersection is found, like an ordinary vanishing point, at 40° to the right. Then on the left a *vertical plane* is obtained to carry the angle of inclination. That is, a vanishing point for 50° on the left is found and is used to indicate the position of a vertical vanishing line. Up this V.L. the angle of inclination is raised precisely as it was in Fig. 68, when a directly ascending plane was treated.

Instead of the angle of inclination (30°) being set up upon the P.V.R., as was done in Fig. 68, it is set up upon what is called a C.V.R. (or central visual ray), which passes from the eye to the V.P. of the direction of the inclination. This C.V.R. we already know as

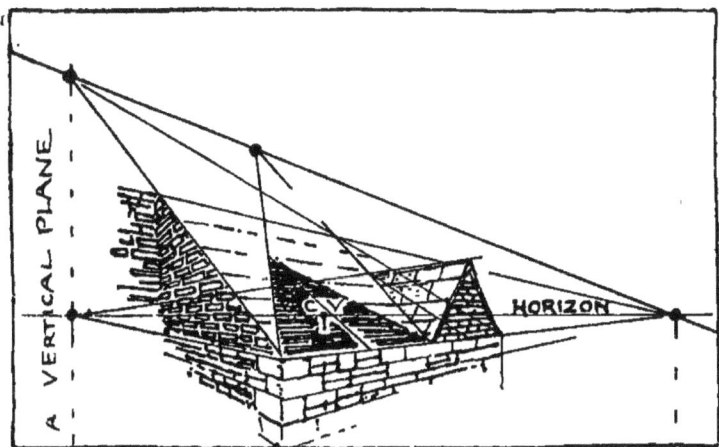

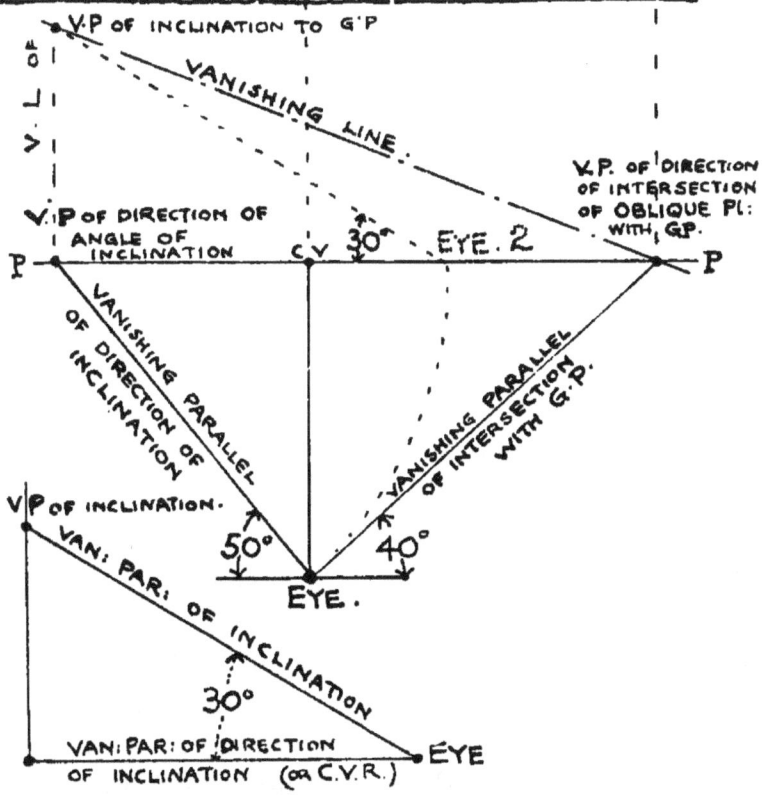

Fig. 70.—The method of finding the V.L. of an oblique plane.

vanishing parallel for 50° on the left.  The P.V.R.
which goes from the eye to C.V. would not serve,
because it has nothing to do with this vertical vanishing
line on the left.

At the foot of Fig. 70 is a triangle showing the
geometric position of affairs, and in the working, in the
middle, the conditions are repeated, partly in dotted
lines, with Eye 2 at the point of the triangle.

By such means we gain two V.P.'s—one of intersection, the other of inclination ; and we know the V.L.
must pass through them.

## 39.

## Laying down the Eye.

So much of the difficulty which will of necessity be
experienced by the student lies in mastering the arrangement of his vanishing lines and the new positions of the
eye, that a few words may be said upon the laying down
of the eye in these new positions.

With Figs. 71, 72, and 73 given here may be taken
Fig. 41, p. 63, where the eye is laid down so as to
serve the horizon.

In all these cases the V.L. acts as a hinge.  We require
the eye in the different positions on our paper because
we have to find points on our vanishing lines by means
of angles drawn from the eye.  Now, the effect of angles
drawn from the eye will vary if the distance of the eye,

which is the point from which they are drawn, varies. If the eye be 10 feet from the V.L. in the picture-plane to which the angles are to be drawn, then an angle at, say, 40° will make a certain distance from the centre C.V. If, however, the distance of the eye be increased by the V.L. not passing through the C.V., then the angle of 40° will not touch the picture-plane till it has expanded further,

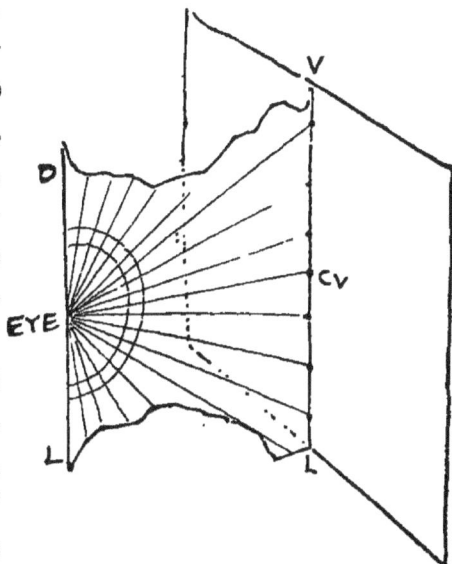

FIG. 71.—The eye in relation to a vertical plane perpendicular to the picture.

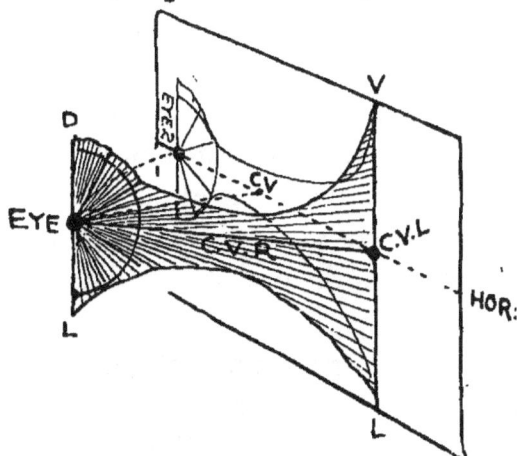

FIG. 72.—The eye in relation to the V.L. of a vertical plane at an angle to the picture, and shown also laid down into the picture-plane.

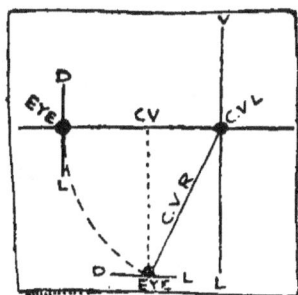

FIG. 73.—The facts of Fig. 72 shown as they appear on the paper when working.

and will thus give a wider measurement than formerly.
Hence it is important that the eye should bear its
correct relation to the various vanishing lines, or the
angles will expand too wide, or not wide enough.

If we take our six possible kinds of vanishing lines, we
find them fall into two groups of three each. The
three V.L.'s which pass through the C.V., the horizon,
the vertical (perpendicular to picture), and the inclined
(perpendicular to picture), will all have the eye only
distant from them the stated distance; that is, if the eye
is said to be 12 feet before the picture, then it will be
12 feet before all these three vanishing lines, because the
12 feet is measured to the C.V. and all these pass through
that point. Hence in each case the distance of the eye
from the V.L. is known and has not to be found. All
that has to be found is its direction in relation to each
V.L. when laid down flat into the picture-plane. For
the eye is really out in front of the picture-plane, and
can thus be at one and the same time in relation
to all possible vanishing lines. But we cannot have
it projecting from our paper—that is impracticable—
so we have to lay it down into the plane of our paper,
and thereon draw the angles we would set off from it.
Necessarily, then, the position of the eye when laid down
for this special purpose will vary.

Now, the student will readily grasp the fact that the
eye is always midway in front of all vanishing lines. No
matter how oblique or long they are, there will always be
a point nearest the eye. This is the *centre of the vanishing line,* or C.V.L. In the case of the three vanishing

lines which pass through the C.V., the centre of them all
is the C.V. The eye is immediately in front of the C.V.,
so that in all three cases the eye when laid down will be
made as far from C.V. as the P.V.R. is long.

These three positions of the eye are shown in Fig. 74,
as also are the three which belong to the three vanishing
lines which do not pass through C.V. In each of these
latter there is a C.V.L. with a line at right angles to it,
C.V.R., running through C.V. and terminating in the eye.

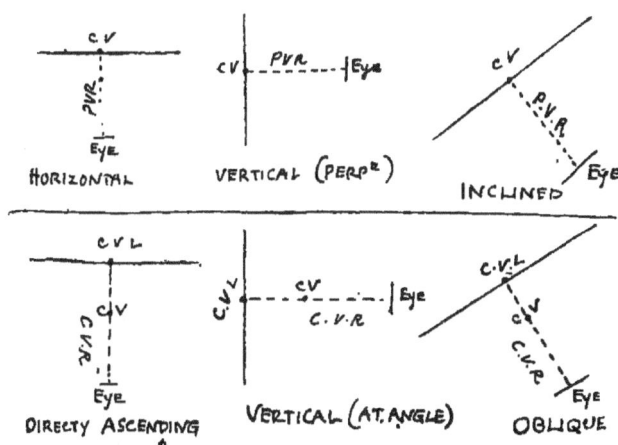

FIG. 74.—The eye in relation to C.V. and C.V.L. in the six different
planes.

It is the length of the C.V.R. which has to be determined. This length is found geometrically; it is the
hypotenuse of a right-angled triangle, the corners of
which are Eye, C.V. and C.V.L. The length of the side
Eye C.V. is always known; it is the distance the eye is
from the picture-plane. The side C.V.–C.V.L. is easily
found by drawing a line from C.V. to the vanishing line

perpendicular to it. Only in the case of oblique vanish-

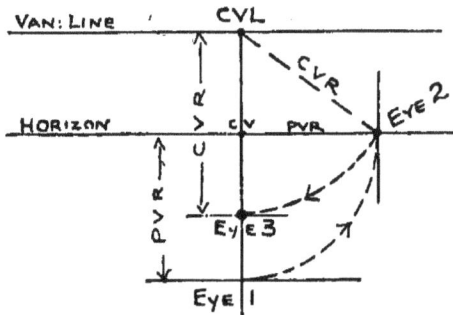

ing lines do we need to geometrically draw this perpendicular; in the case of vertical planes there is always the horizon which is necessarily a line through C.V. perpendicular to any vertical V.L. In the case of directly ascending and descend-

FIG. 75.—Working to obtain eye for V.L. of directly ascending plane. Note that Eye 2 is the position of the eye in relation to a vertical plane through C.V. and C.V.L. C.V.L. is thus a V.P. found on this vertical V.L. from Eye 2.

ing vanishing lines, which are always parallel to the

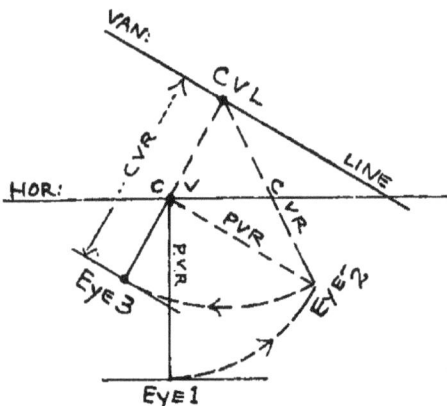

horizon, there is very often a line up through C.V. from the eye which serves for the line required. The working necessary to find the eye in relation to a *vertical plane at an angle* to *the picture* is given in Fig. 73. The working for a *directly ascending* (or descending) and for an *oblique* plane is given in Figs. 75 and 76.

FIG. 76.—Working to obtain the eye in relation to an obliquely ascending plane. The line from C.V. to Eye 2 is drawn parallel to the oblique vanishing line. The arc from Eye 2 to Eye 3 is struck from C.V.L. .

The student is advised to make a paper model on which the triangles used can project, and so that they can be folded down into the plane of the picture.

## 40.

## The Scaffolding of a Problem.

The facts absolutely necessary to the working of a perspective drawing are, as in Fig. 77—

A VANISHING LINE.

A PICTURE-LINE—or several such.

The EYE in relation to the V.L.

(*a*) The vanishing line (V.L.) is the *back* edge of the plane, the picture-line (P.L.) is the *front* edge of the plane.

(*b*) Measurements are found on the V.L. only by setting off angles from the eye.

(*c*) Measurements along the P.L. are actual.

We must note the following facts. The P.L. is often found by measurement downward or upward, to right or left on the C.V. This is when we are told that a particular plane touches the picture-plane, say 5 feet below the eye, above the eye, to the right or left. This does not mean from the "eye" as it occurs in these diagrams, such as Fig. 77. It means 5 feet above *the level* of the eye, and the "eye" in these diagrams is the eye laid

I

down into the picture, and does not give the level
at all—the C.V. does that.

All parallel planes have the same V.L., but each plane
will have its own P.L. See No. 4, Fig. 77.

The P.L. may occur above the V.L., as in No. 2.
Even an ascending plane may commence higher up our
paper than it finishes at its V.L.

The eye may be laid down *above* the V.L., as in No. 3.

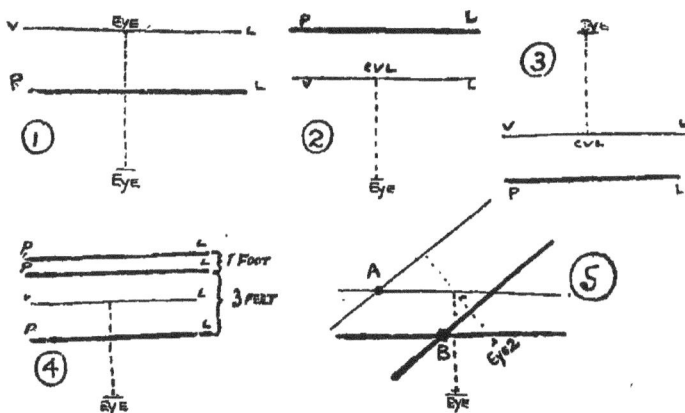

FIG. 77.—The scaffolding.

This makes no difference to the working, and is often a
convenience.

Vanishing lines, being all equally remote, touch each
other where they cross, as A, No. 5; and picture-lines,
being all in the picture-plane, touch where they cross,
as B, No. 5. So that if A and B be joined, the line AB
will represent where the sloping plane cuts through or
intersects the horizontal one.

## 41.

## Finding Picture-lines.

Picture-lines are the l'nes made where planes coming forward from distant vanishing lines strike the picture-plane. It should be noted at the outset that if the planes be perpendicular to the picture, as No. 1, Fig. 78, their picture-lines will be the same distance apart as the planes themselves are—say 2 feet. But if the planes be not perpendicular to the picture, but at an angle to

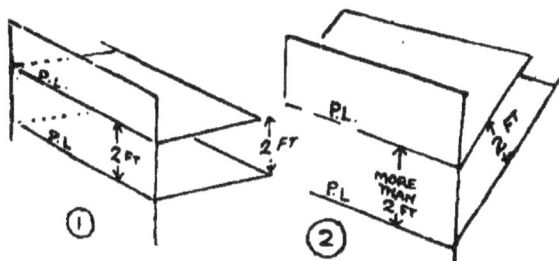

FIG. 78.—Picture-lines declaring (1) and not declaring (2) the distance between two planes.

it, then the distance between their picture-lines will be *greater* than the distance between the planes themselves, as No. 2, Fig. 78. The former case will apply to those planes which have their V.L.'s through the C.V., the latter case to those which have not.

When picture-lines are known to strike the picture-planes so far on right or left, or up or down, the distance is merely measured in actual feet, as in Figs. 79 and 80.

Very often, however, picture-lines, or front edges of planes, occur in unknown positions. A picture-line is known, for instance, to strike the picture-plane somewhere above the eye, or to the right, but exactly where is not *known;* that is, it is not known *in feet,* so cannot be got by measurement. When this is the case, it has to be " found " by perspective methods.

Of course there is always a clue, and the process always

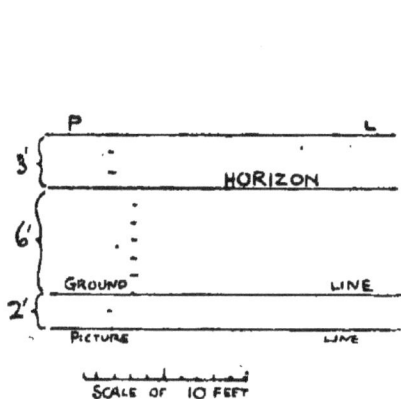

FIG. 79.—Picture-lines found by the distances in feet between them.

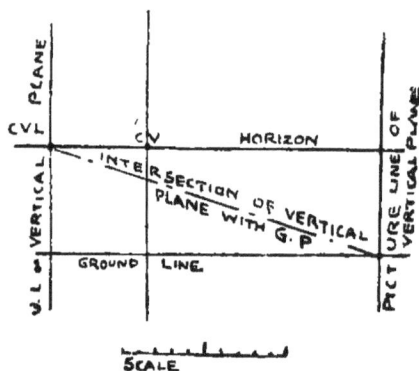

FIG. 80.—A picture-line 14 feet on the right (of C.V.) found simply by measurement.

is to work from one plane, of which we have the picture-line, to another. The business is obliged to start with a *stated* distance, as, for instance, when it is said " the eye is 5 feet above the ground." This statement enables us to at once draw a ground-line (the P.L. of the ground-plane) 5 feet by actual measurement down from C.V. This ground-line is our first line that really belongs to the picture-plane. It is our means of locating all other picture-lines. The first P.L. need not be the ground-line, but it usually is.

The student will realize that in working perspectives we are continually working on shelves, on planes with a space between. We have to get down from one shelf to another, and this we do by using vertical planes, which, of course, cut through both, or all, our shelves. If we can imagine also vertical and sloping shelves, we have to get from one to another by the same means—that is, we use planes which we know cut both.

It is a rule without an exception that we *always get the vanishing lines of planes first*; and the vanishing lines give us the first connections between plane and plane.

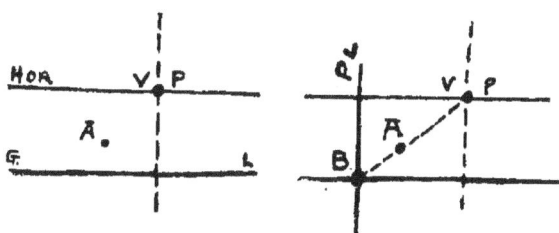

FIG. 81.—A picture-line required of a plane vanishing to the vertical V.L. through V.P., and containing point A on the ground.

Thus, to take the simplest case, suppose in Fig. 81 A is a point on the ground, and a vertical plane from the V.P. comes forward and contains A, where will the picture-line of this vertical plane occur?

Now, the connection between the horizontal plane and the vertical plane is first made by their V.L.'s; that is, by the horizon and the dotted V.L. through the V.P. V.P. is thus the point of connection between the two planes, and is in both. A is also in both, for we are told that it is (1) on the ground, (2) is to be also in a vertical plane, vanishing through V.P. Hence, if we draw down from V.P. through A, we draw a trail of points all of which are in both planes—we draw a line which is in both planes. If we continue this till it cuts the picture-line, we find thereon a point, B, in both the planes, and also in the picture-plane. Vertically through it we draw our picture-line.

It is obvious that this P.L. only applies to point A, or any other point between B and V.P. It is the P.L. of only one of the innumerable planes there may be passing to the same vertical vanishing line.

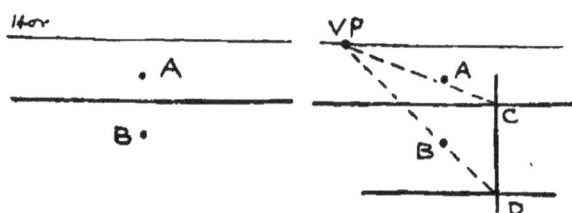

FIG. 82.—Required the P.L. of a horizontal plane containing B, which is a point immediately under point A in the ground-plane.

In working the problem solved on the right side of Fig. 82, we proceed thus. We get down from the plane of the ground to the plane containing B by means of a vertical plane. It does not matter what the vertical plane is, or where it comes from, for since B is immediately under B, vertical planes from any V.P. will serve. We assume a V.P., and draw through A down to C. This is the trace of the vertical plane on the ground, and yields at C the vertical picture-line through C and D. Then we draw through B to D, and at D draw our new P.L. horizontally. Hence the horizontal line through D is the P.L. of a horizontal plane containing B.

Note particularly that if B were *not* immediately under A, then we should have to be told where the V.P. of the vertical plane was; for while a sloping line, such as that from A to B would be, can always be in a vertical plane, it can only be in *one* vertical plane.

We have thus taken a few examples of getting picture-lines, but it will be best to systematically review all the possibilities of the subject. This is done in the next two or three paragraphs.

## 42

## Picture-lines continued.

We are not here concerned with the directions of vanishing lines, nor what their angles are with the planes we commence with. Picture-lines are always drawn parallel to their vanishing lines.

We have already examples in Figs. 81 and 82 of the two forms of working which there are. In Fig. 81 we have the P.L. immediately found after drawing from V.P. through A to B. This is because A is in both the plane we have and the plane we want. In Fig. 82 A is in the plane we have, B is in the plane we want, and we introduce a third plane, so that A can be in the plane we have, and also in the third plane, and B can be in the

third plane and the plane we want. This third plane
is very often a vertical plane, and is generally assumed
for the purpose. In Fig. 82 the picture-line of the
plane of A is connected through the picture-line of the
vertical plane CD with the picture-line of B.

The same process is going on in Figs. 83 and 84. In
both of these cases there is a line AB known to be in an
inclined plane; in Fig. 83 the plane is directly ascending,
in Fig. 84 it is obliquely ascending. All that we have
at the commencement is the horizon and ground-
line, V.L. of our ascending plane, and the line AB.

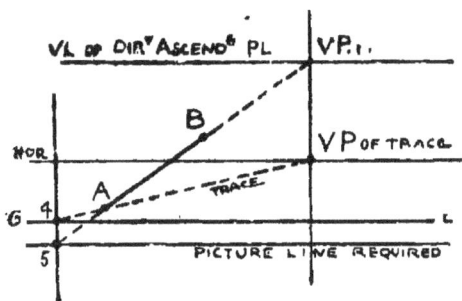

Fig. 83.—Finding the P.L. of a directly ascending plane.

What we do is, we create a line on the ground-plane
*under* the line AB. To do this we find the V.P. of AB,
that is, we continue AB backwards till we obtain V.P.₁.
If now we drop a vertical vanishing line from V.P.₁ down
to cut through the horizon, we obtain a point in the
ground-plane, that is, on the horizon (for the horizon is
part, if only the back edge, of the ground-plane). Hence
the *V.P. of trace* is under V.P.₁ of AB.

What we do next differs slightly according to the way
in which A is related to the ground-plane. Suppose in

Fig. 83 A is known to be on the G.P., and in Fig. 84 its
distance above it is known.

The difference is seen when we draw the trace of the
intersection of our obliging third plane; for if A is on
the ground, we must draw through it. What we do in
both cases is to draw this trace, and then a line over and
under it. In Fig. 83 we draw a line from V.P. of trace
through A till it cuts the ground-line in point 4. Then

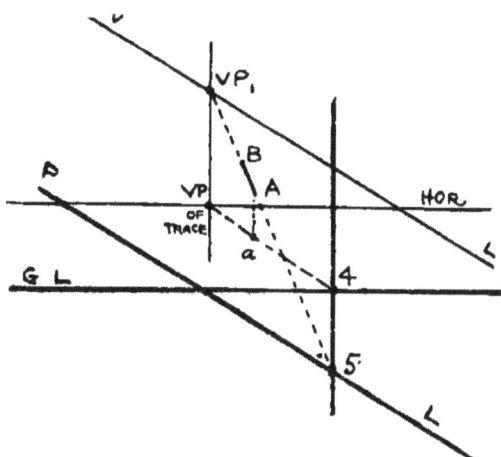

FIG. 84.—Finding the P.L. of an obliquely ascending plane.

from 4 we run up or down a vertical line, which will be
parallel to our vertical V.L. through V.P.$_1$ and V.P. of
trace. We saw that the V.P. of trace was exactly under
the V.P. of AB. If now we draw a line from V.P.$_1$ through
B and A forward till it cuts the new vertical line through
point 4, we find point 5. 5 is just as immediately under
4 as the V.P. of the trace is under V.P.$_1$. The line from
V.P.$_1$ to 5 is certainly in our inclined plane, because it
includes point A, which we know to be in that plane,

and, what is more important, it includes V.P.$_1$, which is also in the inclined plane. Now, point 5 is known to be in the picture-plane, because it is a point on a line through point 4, and 4 is a point on our original picture-line, G.L. Hence if 5 be in the picture-plane, draw P.L. parallel to V.L. through 5, and you have the P.L. of an inclined plane containing AB.

In Fig. 84, where A is so far above the ground-plane, we have to find the seat of A on the ground. This would be done by such lines as those of Fig. 82, where, if we knew that B was 4 feet below A, we should make CD equal 4 feet, and drop a line from A to the line from V.P. to D, finding B wherever it happened to come. B would not in this case be given, but would be found.

## 43.

## Vanishing Points and Lines vanishing to them.

Here are illustrations of the six different kinds of planes which have vanishing lines, and consequently vanishing points on them. In each case the horizon and the eye are given, and the V.L. of the particular plane and the eye in relation to it, with the working shown. The method of getting the vanishing lines, having been treated in paragraphs 32 to 38, is not given here.

In each case a V.P. at 40° on the right of the centre of the V.L., and the corresponding V.P. for a rectangle, that is, at 50° to the left, are given. In the cases of

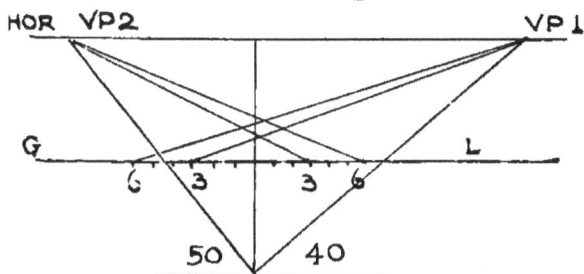

FIG. 85.—A horizontal plane, two lines vanishing at 40° to right, and two at 50° to left.

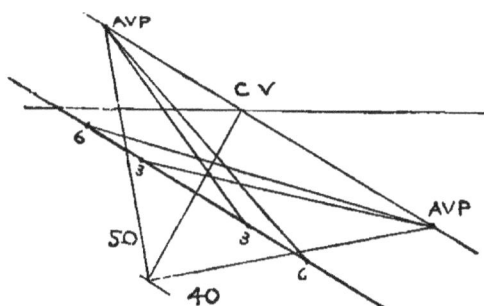

FIG. 86.—An inclined plane.

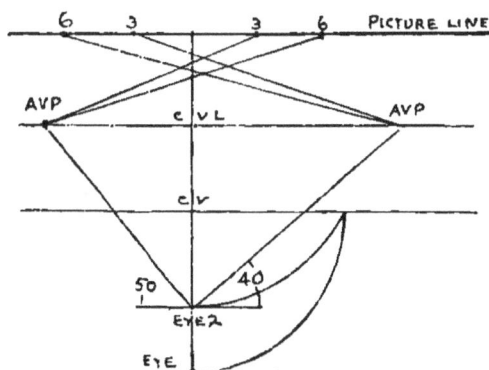

FIG. 87.—A directly-ascending plane.

the vertical planes, *right* and *left* become downward or upward.

The student will observe that we have a front edge
(P.L.) and a back edge (V.L.) of our plane in each case,

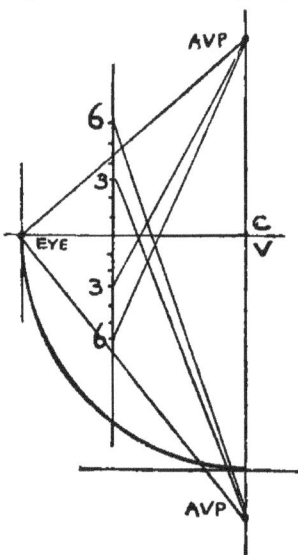

FIG. 88.—A vertical plane (per-
pendioular to picture).

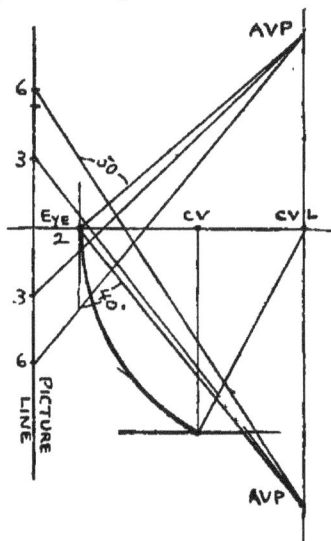

FIG. 89.—A vertical plane (at an
angle to·pioture).

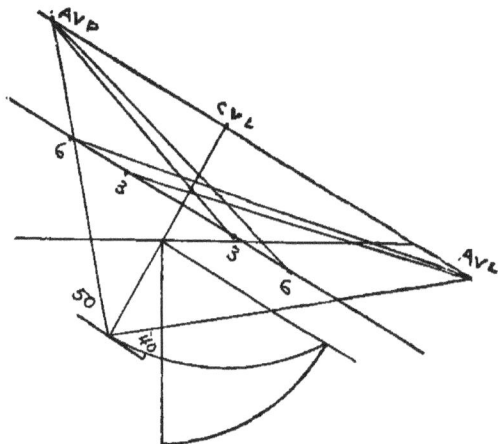

FIG. 90.—An oblique plane.

and that lines pass from the front. edge to the back edge.
'The points of departure on the front edge are 3 feet and

6 feet both on right and left in all cases, and the points of convergence are the same, in description, in all cases.

The crossing of the lines yields a square in perspective, the sides of which are *not*, however, 3 feet, as some might expect, because the picture-line does not pass directly across the lines, but at an angle, and therefore the distance from 3 to 6 is *not* the distance between the lines—but is greater.

The student will surely understand that when the object is not rectangular, and when its position differs from the position of the lines in these examples, the V.P.'s will be found in accordance.

## 44.

## Chess-board in Parallel Perspective shown in all the Planes.

When a rectangle has two of its sides parallel to the picture, the others are necessarily perpendicular to it.

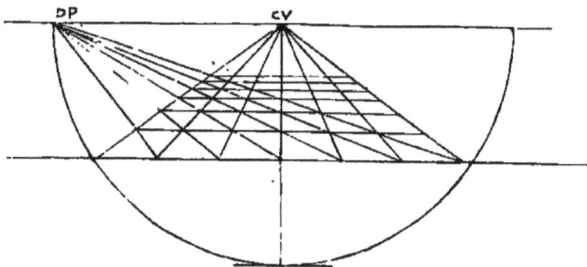

Fig. 91.—A horizontal plane.

We cannot find the V.P. of lines parallel to the picture,

because a line from the eye converging that direction would never meet the V.L., since it is parallel to it.

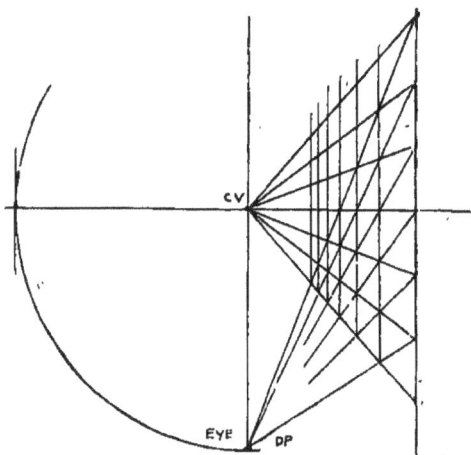

FIG. 92.—A vertical plane (perpendicular to picture).

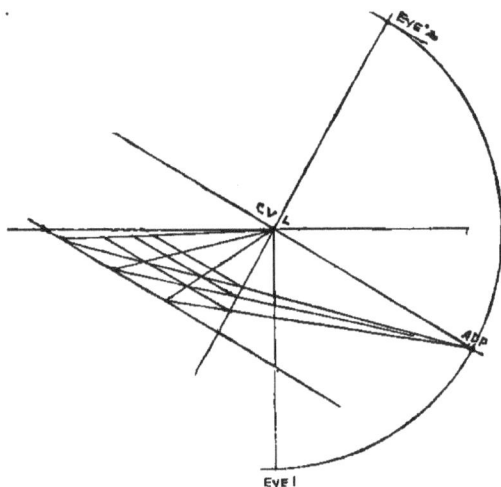

FIG. 93.—An inclined plane (perpendicular to picture).

The other lines perpendicular to the picture vanishing straight in, and a parallel from the eye converging this

direction, will strike the V.L. either at C.V. or C.V.L.
The three V.L.'s passing
through the C.V. have it as
their centre; the three V.L.'s
which do not pass through
C.V. have C.V.L.'s instead.

The distances between the
lines are carried back by the
lines going to C.V. or C.V.L.,
so that distances between the
lines are also truly recorded
on the picture-line. In all
cases, too, we have our distance
point (D.P.) as measuring
point.

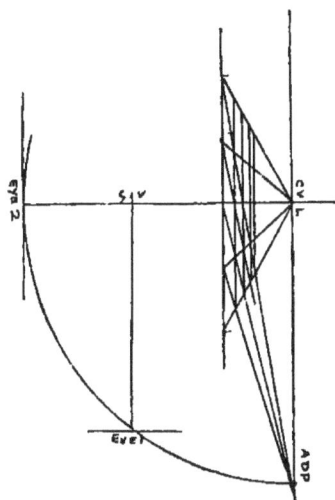

FIG. 94.—A vertical plane (at
angle to picture).

The next paragraph is devoted to converting such

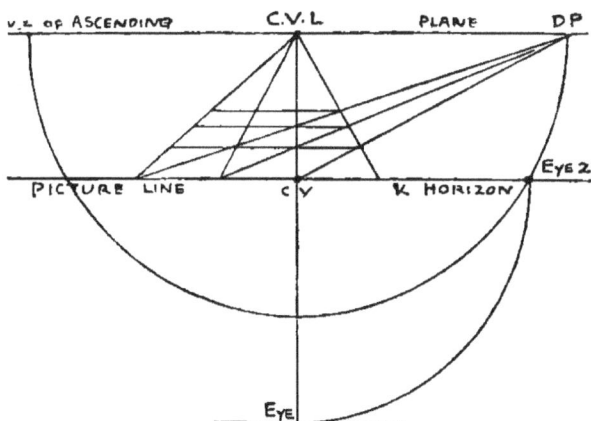

FIG. 95.—A directly ascending plane.

squares as these conditions yield into cubes; that is, we

have here two of the three dimensions of space, we have to get the third.

What should be chiefly observed with these planes and chess-boards is that they yield the *lines of heights* we so often use. Thus Figs. 92 or 94 give us lines of height measuring distances above and below the ground-plane.

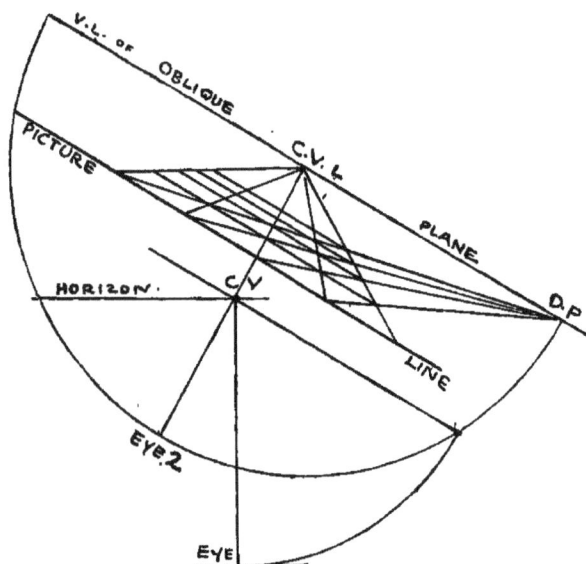

FIG. 96.—An oblique plane.

The term "parallel perspective" is not usually applied to any but the first of these examples. The various instances are given here because the comparison is instructive.

## 45.

## Perpendiculars to the Six Different Planes.

Three of our planes are perpendicular to the picture; the three, that is, which have their V.L.'s through C.V. Perpendiculars to these planes will be lines parallel to the picture, and will consequently be drawn with the **T** or set square at right angles to the picture-lines of the planes.

Perpendiculars to planes the V.L.'s of which do not

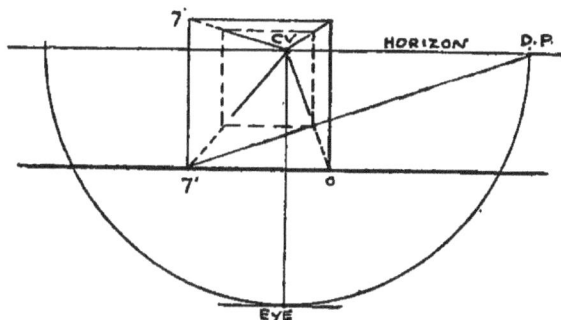

Fig. 97.—A square of 7 feet sides lying in ground-plane converted into a cube by vertical lines 7 feet high obtained by a line of heights, or a portion of such a chess-board as that in Fig. 92.

pass through C.V.—planes, that is, inclined to the picture —are also inclined to the picture, and have to be vanished and measured.

The V.P. of the perpendiculars is readily found. It always has a vanishing parallel at right angles to the parallel, by which the direction or inclination of the plane is obtained. Thus in Fig. 100, V.L. is the vanishing line of a vertical plane at 48° to the left. The angle 48° is set off from the eye, and at right angles to it a

K

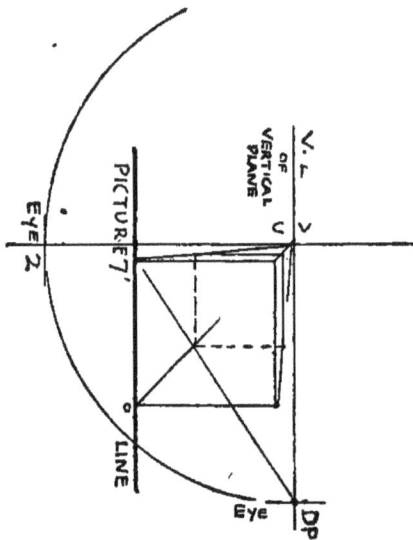

FIG. 98.—A cube of 7 feet in a vertical plane perpendicular to the picture. Its height, which is horizontal, is found by a portion of a chess-board similar to that of Fig. 91.

parallel, 42°. to right, yields the V.P. of perpendiculars to all planes vanishing in the V.L. at 48° to left. Hence, the base of a cube being found in the plane, the "upright" edges of the cube will vanish to the V.P. on the right. These "upright" lines are in horizontal planes vanishing on to the horizon. To measure the lines, we draw a picture-line, and measure by the M.P. for the V.P. of perpendiculars.

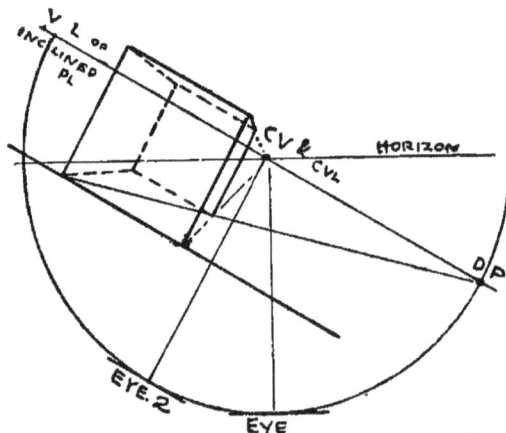

FIG. 99.—A cube of 7 feet standing on an inclined plane perpendicular to the picture. Its height is found by the chess-boards of other inclined planes perpendicular to picture, and also perpendicular to the plane of the base.

In Fig. 101, the ascending plane is at 31° to the horizontal. The vanishing parallel, 31°, is set up at the D.P. (or Eye 2), and the vanishing parallel of perpendiculars at 59° is set down. The two vanishing parallels are at 90° to one another.

Instead of a cube with base lying in an oblique plane, the diagram, Fig. 102, gives only one perpendicular, AB. If A be supposed to be one of the corners of a square, then AB would be one of the edges which would convert that square into a cube. The diagram appears very complicated, but the lines are not those necessary for obtaining a perpendicular to an oblique plane, but are *all* the V.L.'s and P.L.'s which belong to a rectangular object with base in an oblique plane.

Slanting across the top is the V.L. of the oblique plane, with A.V.P. at one end and V.P. at the other. A.V.P. is the *V.P. of inclination* obtained by a vanishing parallel from eye 2, which will be found on the horizon. The vanishing parallel is not drawn, because there is great danger of confusion; it passed near point B. Under A.V.P. is C.V.L., occurring where the *V.P. of the direction of the inclination* was found. It was through this V.P. that the vertical V.L. was drawn. Of this V.L., C.V.L. is the centre.

Eye 2 is the position of the eye in relation to this vertical V.L., and eye 2 gave us the inclination of A.V.P. upward (that of the plane), and gives the declination downwards of A.V.P.$_2$, which is the V.P. of the perpendiculars.

We do not hesitate to set up *lines of height* perpen-

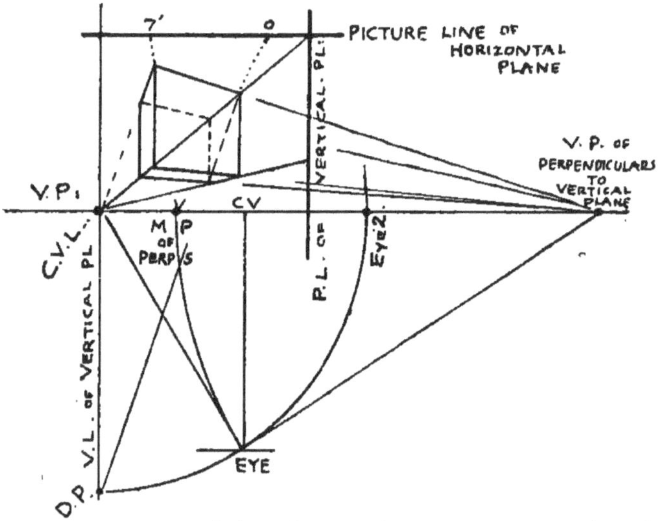

FIG. 100 —A cube with base in a vertical plane at 48° to the left.

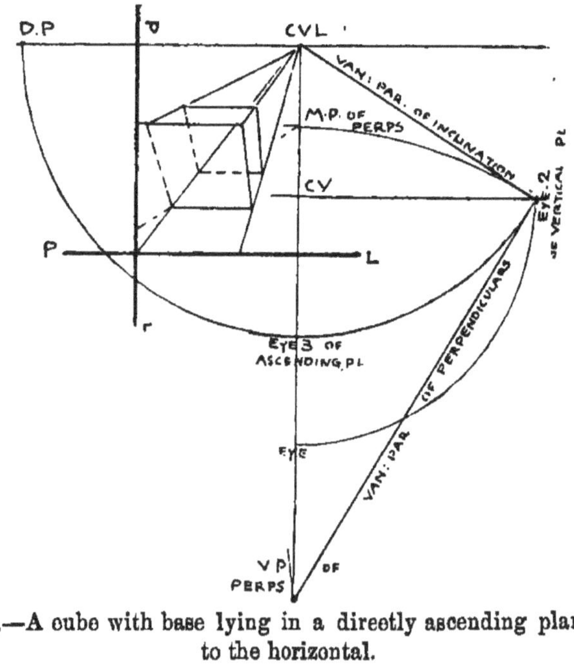

FIG. 101.—A cube with base lying in a directly ascending plane at 31°
to the horizontal.

dicular to the horizon as often as we like. Could we set up similar lines of height perpendicular to the V.L. of our oblique plane? No, because the perpendiculars to any planes whose V.L.'s do not pass through C.V. vanish; that is, they do not remain parallel to the picture, and consequently lines of height on the picture will not be parallel to them, and the measurements on both will not correspond.

A line drawn up through the V.L. of the oblique plane, *and containing A.V.P.$_2$* (Fig. 102), will be the V.L. of a plane perpendicular to the oblique plane. The upstanding sides of a cube lying on such an oblique plane as this will vanish in V.L.'s of different planes, oblique to the picture-plane, but perpendicular to the oblique plane on which the object stands. These V.L.'s are given in Fig. 102.

Thus, the base and top of the cube would vanish on the V.L. through A.V.P.$_1$ and V.P. If, *and only if*, the sides of the base and top vanished to A.V.P. and V.P., the square sides of the cube would vanish upon the V.L.'s through A.V.P. and A.V.P.$_2$, and V.P. and A.V.P.$_2$. In short, the large triangle is made up of the three V.L.'s for the cube. AB may be considered one of the corners.

There are three V.L.'s passing upward from A.V.P.$_2$. One, the central, is through C.V. and C.V.L. AB is in all these three planes, and it could be in any others that contained A.V.P.$_2$.

The small triangle of thick lines is the P.L.'s of the V.L.'s. The lowest is the P.L. for the V.L. through

A.V.P. and V.P., and its position is assumed in this case. The uppermost P.L. which belongs to the lowest V.L. was found by drawing a dotted line from V.P. through A till the first P.L. was cut.

The others were found in the same way, the dotted lines used being shown.

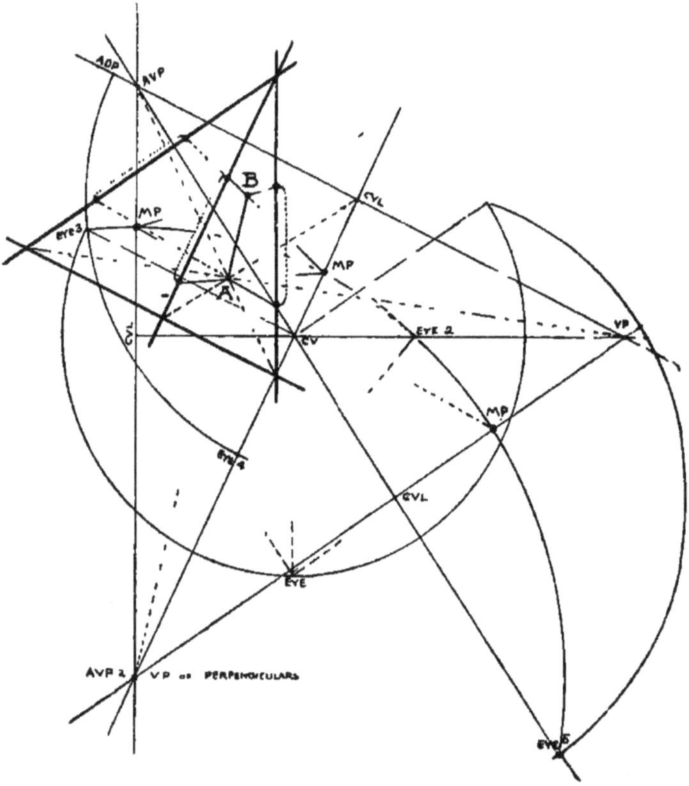

Fig. 102.—The V.L.'s, P.L.'s, and different positions of the eye belonging to a rectangular object (as a cube) with one face in an oblique plane.

AB is measured on all the three perpendicular planes. Note that the eyes in relation to each (eye 2, 3, and 5) all occur on the same arc, as also do the three M.P.'s.

The perpendiculars to the seventh plane, parallel to the picture, are horizontal, and vanish all of them to C.V. In fact, they are the lines running in to the picture in parallel perspective. Nothing more of them need be said.

We have thus completed the system. The subsequent

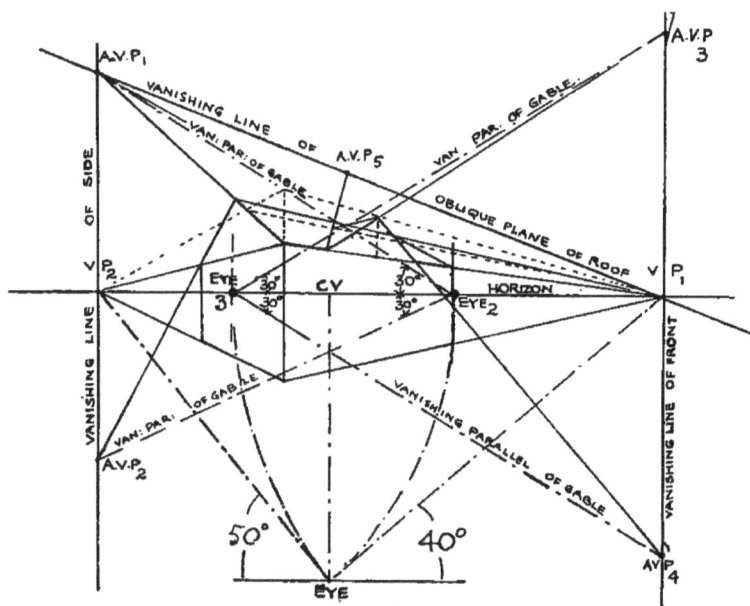

Fig. 103.

solutions of examination questions supply sufficient examples.

Fig. 103 is added to show the way the working lines of the geometry, by which the V.L.'s, M.P.'s, and eyes are obtained, conspire to obliterate the picture and confuse the student!

## 46.

## A Few Hints to Draughtsmen.

Although architects use their own method of *architects'*
perspective, which has been explained in paragraph 24,
yet a very great deal, nay, possibly as much and more,
can be done by the ordinary method which forms the
chief part of this book. To gain the advantage, however,
two or three matters must be observed. First, one should

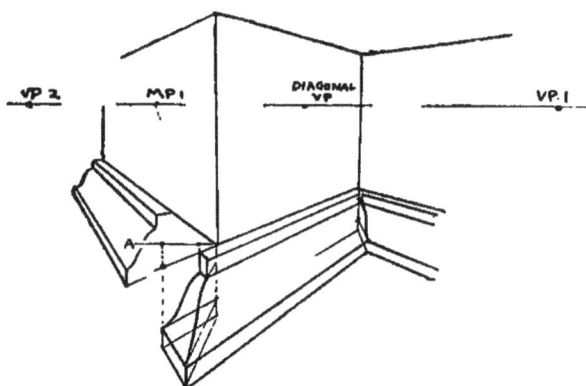

FIG. 104.—Mouldings worked out before the picture. The picture-line is
the short line through A. Measurements are cast forward from it
by means of M.P.1.

work before the picture—that is, let the drawing come
down below (in front of) the ground-line. How one fears
to do this! and yet it is as theoretically sound as any
other action in perspective.

In Fig. 104 the mouldings are obtained by working
forward.

Our second injunction, to keep our lines short, is acted

upon in Fig. 104 also. Thus the ground-line is the line from A to the right, just of sufficient length.

Again, the V.P. of diagonals saves a large number of lines, and is also used in Fig. 104.

If, lastly, we push our picture-plane back or drag it forward so as not to have to use a long ground-line where a short one can readily be found, we save a good deal of line. In the Jesuit's perspective there is a street scene

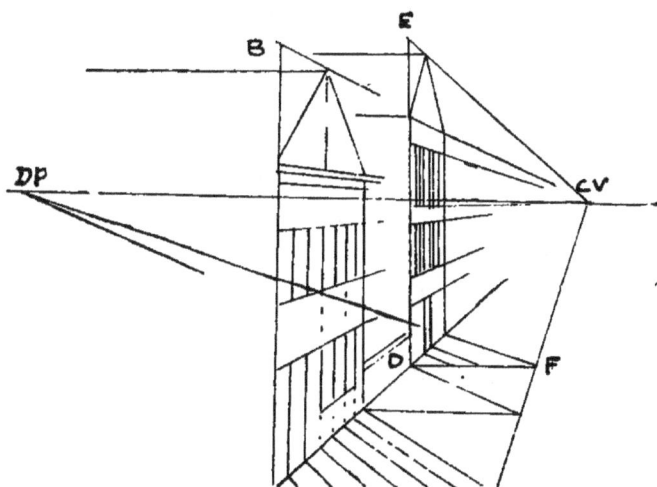

Fig. 105.—A new ground-line for each house.

in which the fronts of the houses are measured from new ground-lines. Fig. 105 is practically a sketch reproduction of part of the Jesuit's diagram. He covers his street with a chess-board, as he does everything. This gives him a number of horizontal lines carrying the scale of the ground-line back. He tells us to raise a new line of "elevations" at the angle of every house. That this method may be made considerable use of with advantage there can be no doubt. Students, as distinguished from draughtsmen, should, however, beware of these rather fanciful methods, which, sound as they really are, sometimes appear to students as short cuts and ready methods. Only persons really at home in perspective can avail themselves of these facilities.

The practical draughtsman has sometimes a difficulty in drawing to the V.P.'s, which occur perhaps at a great distance on right and left. An instrument called a centrolinead, which is illustrated in Peter Nicholson, is sometimes used. The best way, after all, of getting lines from a distant V.P. is to have a long straight edge. A ready method for finding the directions of lines to inaccessible V.P.'s is given on p. 47, and is, I believe, as practicably useful as any.

The following is the usual geometric method. The horizon and vanishing parallel are given (Fig. 106), and a line is required from point A to its V.P., which occurs where the parallel touches the horizon. Draw any two parallel lines BC and DE. Join CA and BA. Draw EF

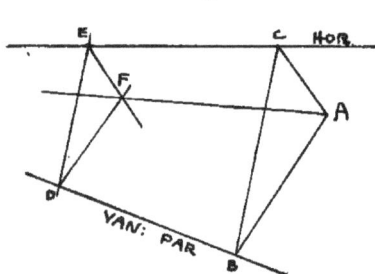

Fig. 106.—Geometric method of drawing an inaccessible V.P.

parallel to CA and DF parallel to BA. The line required will run through point F.

As a rule, as the reader is by this time aware, four or five points usually occur on the horizon. They are V.P.$_1$, M.P.$_2$, M.P.$_1$, V.P.$_2$, or a V.P. of diagonals

may occur somewhere between the two M.P.'s. There
is no ready method of finding these except for lines
vanishing at 45° to right and left—to the D.P.'s, in
fact. Now the vanishing parallel for V.P. 45° (D.P.) is
the diagonal of a square whose side is the length of
eye from C.V. Now the diagonal of a square is to
the side as 17 is to 12. If, then, a draughtsman set off
the two angles of his object from A (Fig. 107), and
determine to have his V.P.'s as D.P. and D.P., then
he knows that halfway between them will be C.V., his
V.P. of diagonals, and that if from C.V. to D.P. be
12 inches, from D.P. to M.P. will be 17. These figures

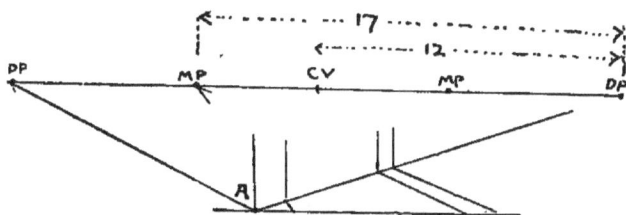

FIG. 107.—The M.P.'s and V.P. of diagonals for lines vanishing at
45° found without geometry.

happen to be very suitable, because 12 inches is a
measurement he is much accustomed to using.

Most designers begin their designs by making a
perspective sketch. Now, if he can find by perspective
what sizes actually will yield the aspect obtained in
his sketch, he can proceed with his elevations with
more confidence. His sketch perspective being quite
a freehand performance, the more neatly he draws it
the surer will be his final results. Now, if we have
the two V.P.'s *and the C.V.* we can get the eye and so

get everything; for a semicircle upon the horizon from
V.P. to V.P. will have a line vertically from the horizon
from C.V. down to eye (see Figs. 161 and 197). But the
C.V. is not a point suggested at all by the sketch; in
fact, unless the object either vanishes to it or to 45°
right and left, the C.V. has nothing to do with the work.

However, the V.P. of diagonals has, and nothing is
easier than, if the sketch is neatly drawn, to draw a
horizontal square as a part of the design, and find the
V.P. of its diagonal. This is done in Fig. 108. There
the sketch of the fireplace was made entirely before any
V.P. or other lines were drawn. Then V.P.$_1$, V.P.$_2$, and
V.P. of diagonals were found merely by continuing the
lines of the design to a conclusion.

We can find our eye if we have our V.P. of diagonals.
For V.P. of diagonals has a vanishing parallel which
bisects the angle between the vanishing parallels of V.P.$_1$
and V.P.$_2$. That angle is 90°. Settling our attention on
the left side of the drawing, we know that the vanishing
parallel of V.P. of diagonals will make half 90°, that
is 45°, with the vanishing parallel of V.P.$_1$. The student
will recollect that segments of circles have the property
of retaining within their curve a certain angle. For
example, a semicircle is the segment containing 90°,
which means that two lines from the extremities of the
diameter meeting on any point of the circumference will
make between themselves 90°. We know our vanishing
parallels for V.P.$_1$ and V.P.$_2$ must have 90° between
them, so we know that the eye will occur on a semicircle from V.P.$_1$ to V.P.$_2$; we get this semicircle. Apply-

ing the same principle, we know that two lines from the
ends of the line from V.P.$_1$ to V.P. diagonals, that is, from
those two points, will contain 45°. If, then, on that line
as a chord we form a segment of 45°, it will give us the
only possible positions these lines can occupy. We make
our segment thus. Bisect line V.P.$_1$, V.P. diagonals, and
draw a perpendicular. From V.P.$_1$ set down from line

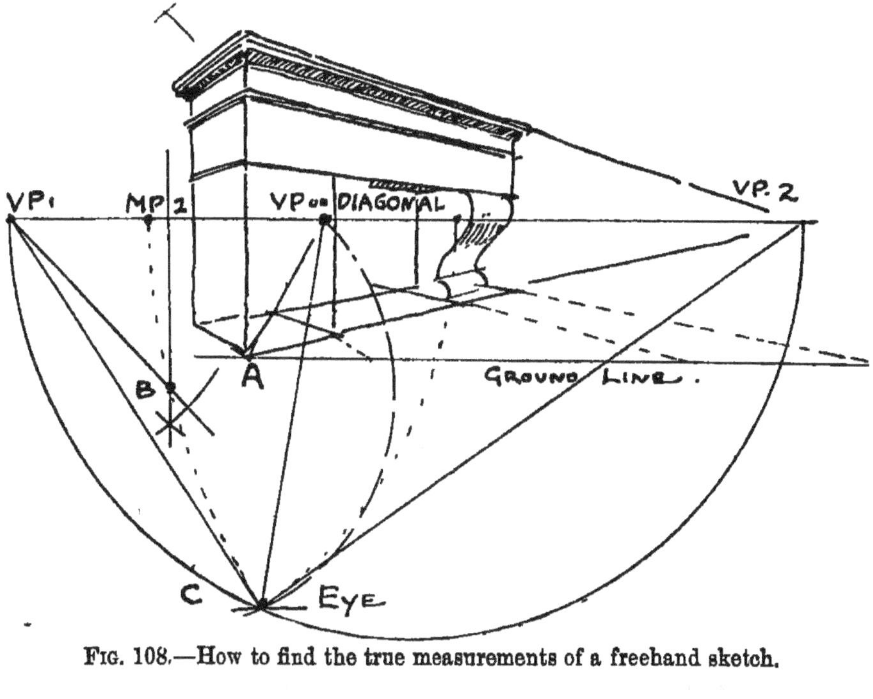

FIG. 108.—How to find the true measurements of a freehand sketch.

V.P.$_1$, V.P. diagonals, 45° (which is the difference between
the angle we want, 45° and 90°), and where this angle
cuts the perpendicular will be the centre of the circle
yielding the segment. B is, therefore, the centre. Draw
the segment, or such a part of it as is needed to cut the
semicircle. Thus C is found, and C is the position of the

eye. From it the M.P.'s can be found, and a ground-line being assumed through A, measurements can be taken.

## 47.

## Projection of Shadows.

Shadows are of two kinds—those cast by an artificial light, and those cast by the sun. In both cases the shadows of lines have direction, that is, they are drawn from points. Those, however, cast by artificial light *radiate* from a point, or points, near them, while those cast by the sun *vanish* to vanishing points. If, then,

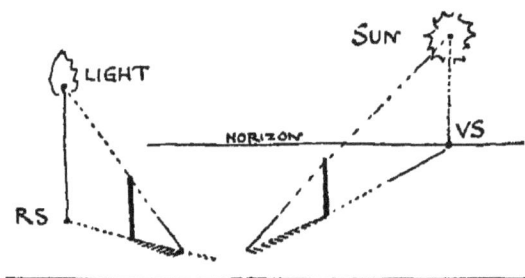

FIG. 109.—Shadows cast by artificial light and by the sun.

several vertical lines have their shadows cast on the ground by an artificial light, the shadows will be seen to converge and meet at a point on the ground, which may be called a *point of radiation of shadow*, R.S., as in Fig. 109. But if the shadows be cast by the sun, they will vanish to a vanishing point of shadow, V.S., as in Fig. 109.

The reason for this difference is found in the fact that an artificial light is so near to, so much "among," the object it casts the shadow of, that its rays are spread around on every side, while the sun's rays are held to be parallel, as if the sun were a luminary which cast its rays in but one direction only. Fig. 110 shows how little the all-round radiation from the sun affects the earth.

In all cases there are three factors—

1. The luminary.
2. The line.
3. The plane the shadow falls on.

Both the luminary and the line must be related to the

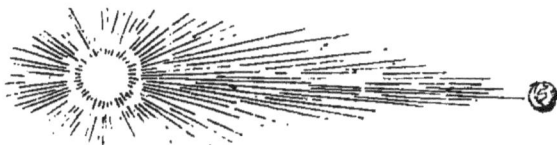

Fig. 110.—The sun's rays parallel as far as the earth is concerned.

plane. They must be known to be over some point on the plane. If this relationship is definitely given or can be definitely found, or inferred, we can work a true sciography of the shadows; if we have to guess the relationship, the problem is one of sketching or drawing, not of working. A reference again to Fig. 109 will show that for each shadow we have a line on the plane receiving the shadow, and that this line contains *three points*. Of these three points one represents the luminary. This point is either R.S. or V.S. In both these cases the point is found by dropping a line from the luminary in harmony with the direction of the line whose shadow is required.

If in the second case it touches the ground-plane so far back as the horizon, it is because the sun is that far back. Another of the three points is the foot of the line; the third is the limit of the shadow, or the shadow of the head of the line on the ground.

All shadows are found as these are found. The only variation is that caused by the lines not being vertical, the plane not being level, and so on.

We take first the shadows cast by an artificial light.

## 48.

### Shadows cast by Artificial Light.

The following rule applies in every case:—

*Draw a line from the light to the plane receiving shade parallel (in fact) to the line whose shadow is required. The point found on the plane is the R.S. (point of radiation of the shadow).*

*Shadows cast on the ground-plane (by artificial light).*

The working applies, of course, to any *horizontal* plane, and not merely to the ground-plane.

Lines can occur in three positions in relation to the ground. These are illustrated in Fig. 111.

A is vertical to G.P., and has no V.P.

B is parallel to G.P. or horizontal, and will vanish somewhere on the horizon, say to V.P.₁.

C is inclined to the G.P. It vanished to A.V.P., which

is a point in a vertical vanishing line cutting the horizon in V.P.$_2$.

Applying the rule to each case, we obtain the points of radiation of the shadows, R.S. 1, 2, and 3, as follows :

For the R.S. of line A we draw a line parallel *in fact* to the line. Line A does not vanish, so a line parallel to it will not vanish. Hence we draw vertically down to R.S., a point which we must already have had given us, because the relation between luminary and plane of shadow must always be obtainable.

R.S.$_1$ being then the radiating point of the shadow of A,

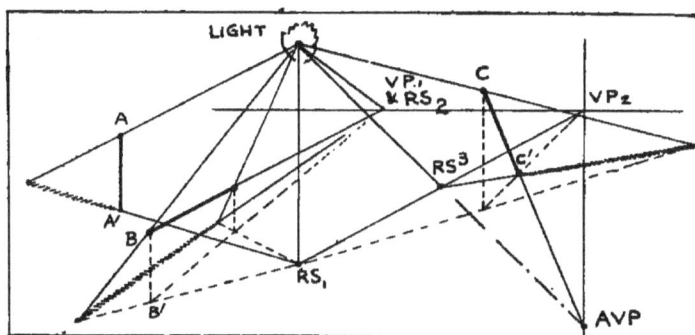

Fig. 111.—Shadows cast on the ground-plane.

we draw from R.S.$_1$ through the seat of A on the plane, and limit the line by a ray from the light.

B is a horizontal line vanishing to V.P.$_1$. To find R.S., draw a line from the light parallel *in fact* to B. This new line from the light, being parallel to B, will vanish with it. Where it cuts the plane of shadow, that is, the ground, will be the R.S. But it does not cut the ground till the horizon is reached. Nevertheless the horizon is

L

part of the ground, and so the V.P. and the R.S. coincide:
R.S.$_2$ is then the radiating point of the shadow of B.   B
being above the ground, we have to get down to the
ground by shadowing BB′ as we did line A.   The shadow
of B is limited by the rays from the light.   The shadow
of the vertical at the far end of B being also thrown, and
coinciding with the result just obtained, proves its
correctness.

Line C is inclined to the ground. It vanishes to A.V.P.,
so that a line from the light parallel in fact to C will
have to vanish there too.   To find where this line
actually touches the ground we have to employ V.P$_2$,
which is the point on the horizon immediately above
A.V.P. Now R.S.$_1$ is the seat of the light on the ground;
a line, therefore, from R.S.$_1$ to V.P.$_2$ will lie on the ground-
plane, and will be such as will be cut by the line from the
light to A.V.P.   This gives R.S.$_3$, which is the radiating
point of the shadow of the line C.   We draw from R.S.$_3$
through the end of C, namely C′., and limit the shadow
by a ray from the light.   To prove this, point C is treated
like point A, and its shadow
is found by a vertical line.
This is shown in dotted lines.

*Examples of shadows cast
on the ground-plane.*

The student knows that
a cube has three vanishing
points, or rather, that the
twelve lines of a cube vanish

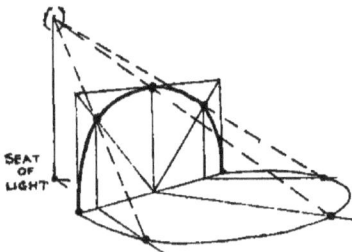

Fig. 112.—A semicircle standing
vertically, its shadow cast by an
artificial light.

in sets of 4 each to 3 different V.P.'s.   In Fig. 115 these

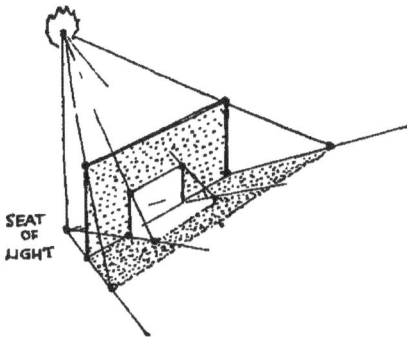

FIG. 113.—An upright rectangular form with its shadow on the ground

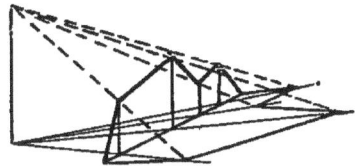

FIG. 114.—A zigzag line and its shadow obtained by verticals from the angles to the ground.

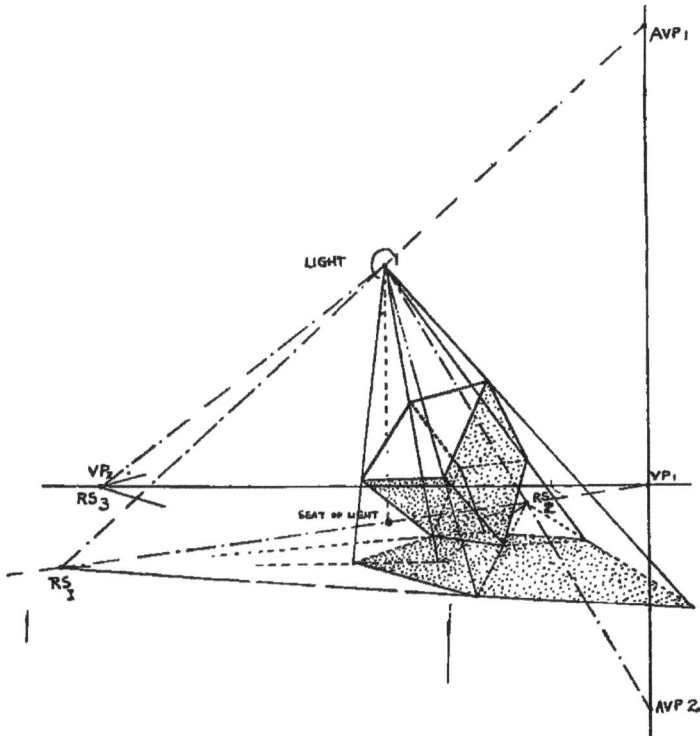

FIG. 115.—A cube in oblique perspective, with its shadow cast from an artificial light.

three V.P.'s are A.V.P$_1$, A.V.P.$_2$, and V.P.$_2$. The student will see in the diagram four chain-lines ; one runs from V.P.$_1$ to R.S.$_1$. It is the trace of a vertical plane vanishing in the vertical V.L. through V.P.$_1$, and it contains A.V.P.$_1$, A.V.P.$_2$, and the light. This trace shows where lines from the light to A.V.P.$_1$ and A.V.P.$_2$ will cut the ground.

The other three chain-lines are those by which the points of radiation of shadow are obtained. Thus one from the light to V.P.$_2$ is parallel (in fact) to the 4 lines of the cube which vanish at V.P.$_2$. The shadows of these lines can readily be traced, also vanishing to V.P.$_2$.

The remaining lines of the cubes vanish either to A.V.P.$_1$ or A.V.P.$_2$. A chain-line from the light through A.V.P.$_1$ gives R.S.$_1$, the radiating point of the shadows of the lines vanishing to A.V.P.$_1$. Our last chain-line from the light to A.V.P.$_2$ gives R.S.$_2$ the point to which the shadows of the lines vanishing to A.V.P.$_2$ radiate. The process is precisely the same as that of C, Fig. 111.

## 49.

## Shadows cast upon a Vertical Plane by Artificial Light.

In this example the vertical plane is perpendicular to the picture, and therefore lines perpendicular to *it* are

parallel to the picture and do not vanish. The light is
somewhat to the left. To obtain the R.S. of AB, we draw
from the light a line parallel to AB till it cuts the wall.
This it will do on the chain-line which is formed by the
line from the seat of the light on the ground to the intersection of wall and ground. R.S.$_1$ is thus the point to
which the shadow of AB will radiate. B' is therefore

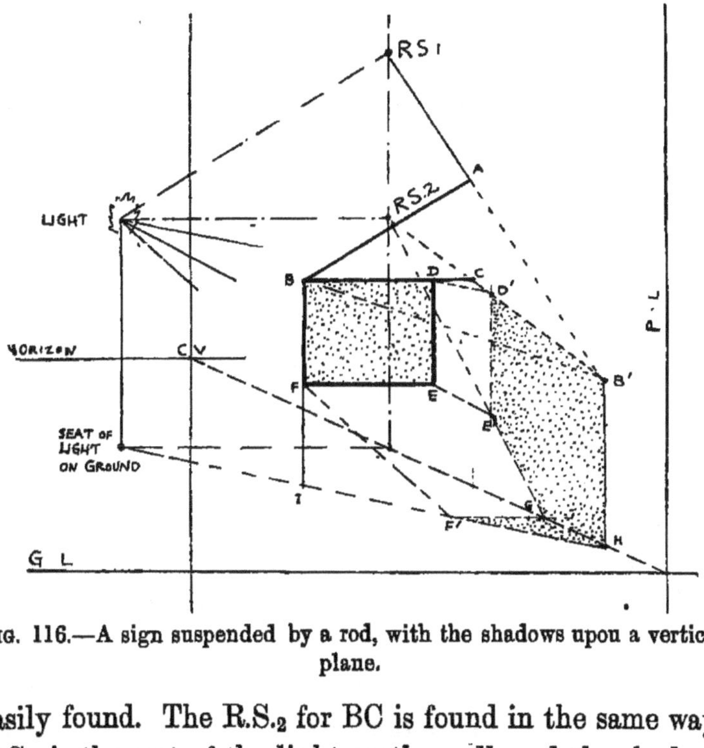

Fig. 116.—A sign suspended by a rod, with the shadows upon a vertical
plane.

easily found. The R.S.$_2$ for BC is found in the same way.
R.S.$_2$ is the seat of the light on the wall, and the shadows
of all perpendiculars to the wall, such as BC and FE are,
will radiate to it. The shadows of BF and DE will be
drawn vertically downward, since a line from the light
parallel to them would never reach the wall.

Part of the shadow falls on the ground, the edges affected being part of BF and part of FE. The shadow of FE on the ground will have no R.S., because a line parallel to FE from the light would never touch the ground. BF will be treated as verticals always are; its shadow will radiate to the seat of the light on the ground.

The student should note that BF and FE afford instances of shadows commencing on the ground, F'H and F'G, and running up the wall.

## 50.

## Shadow cast on an Oblique Plane by Artificial Light.

In this figure we have instances of four kinds of lines throwing their shadows on an oblique plane. The lines are the vertical, the horizontal to left and to right, and a line DE at an angle different from that of the plane.

The shadows are found precisely as in former cases.

The vertical AA throws a shadow upon the sloping back. We draw from the light a line parallel to AA— that is vertically—and where it cuts the sloping plane is the point from which the shadow radiates. To obtain this point we draw a line, M, through the seat of the light on the G.P., to cut the intersection of ground with the sloping plane—this intersection runs through A to

V.P.$_2$. We then bring down a line from A.V.P.$_1$, which represents the inclination of the plane perpendicular to

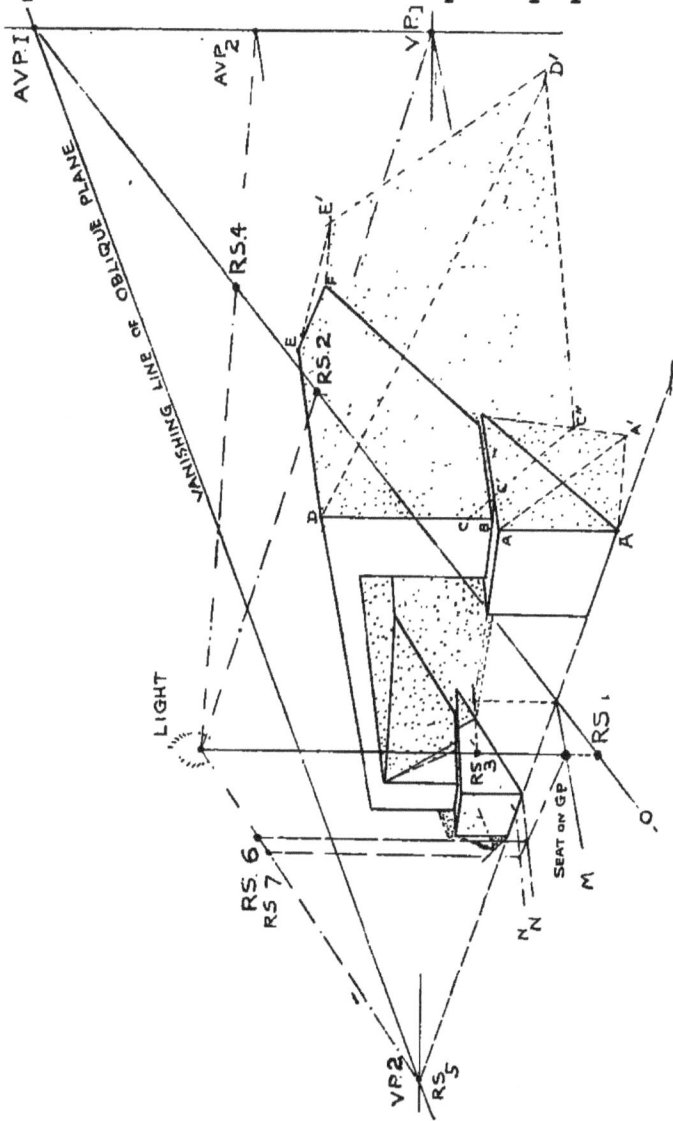

Fig. 117.—A structure built against a sloping bank which is part of an oblique plane. The shadows are cast from a light, the height of which and seat on ground-plan are given.

the intersection, and it comes down towards O. We

drop our light down to this line and gain R.S.$_1$, which is the point to which the shadow of AA will radiate. We thus easily obtain AA', the shadow.

The shadow of the vertical BD on the narrow upper surface of the plinth requires one to find the seat of the light on the plane of the top of the plinth. By using V.P.$_2$ and then V.P.$_1$ we transfer this level round till we get R.S.$_3$. BC' radiates from R.S.$_3$. AC' has its R.S. found by drawing from the light a line parallel to it—that is, to its V.P., which is V.P.$_1$. R.S.$_2$ is thus obtained. DE is dealt with in precisely the same way.

For the shadow of the horizontal lintel on the wall inside, we have to find the seat of the light on the plane of the wall and the plane of the plinth, R.S. 6 and 7. The one point will not serve the two planes.

### 51.

## Shadows of and on Cylinders, etc., thrown by Artificial Light.

A circle has no sides, and hence has to be treated in a special manner. We obtain the perspective shape of a circle by assuming points in its circumference and drawing the curve through them by freehand. We find the shadows in somewhat the same way. The shadow of a circle itself is not difficult to find; one merely gets shadows of the assumed points, as in Fig. 112. But when the object is a cylinder, or a hollow cylinder, we have to use the method shown in the small sketch on Fig. 118.

There L is the light and S the seat of it on the plane of the circle. A line from S across the mouth of the hollow cylinder, or tube, will give points A and B. A plane can readily be imagined which will contain the line SL and the line SB. Such a plane would produce the line BC on the inside of the tube, because we are

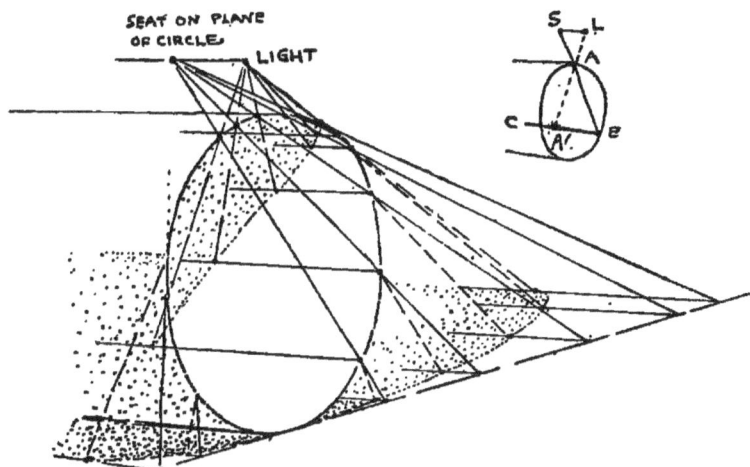

Fig. 118.—The method of obtaining shadows of cylinders.

careful to make SL follow the direction of the tube. A ray from L through A will strike BC in A', and A' will be the shadow of A.

In the larger diagram lines are drawn across the mouth of the tube from the seat of the light; they are also carried down to the ground. They yield points from which can be drawn lines in the tube and on the ground, such that rays from the reciprocal points above will record on them the shadow positions of those points.

## 52.

## Shadows cast by the Sun.

The process of obtaining these shadows differs from those thrown by artificial light, in that the sun being infinitely distant, and its rays being parallel, the VP of the sun's rays becomes the luminary. This use of a vanishing point throws the working at once back to the region of vanishing points and vanishing lines. Our shadows no longer radiate to a point on the plane near the lines they belong to, but to a point on the farthest edge of the plane—its VL.

Our rule becomes—

*Draw through the V.P. of sun's rays and the V.P. of the line whose shadow is required. Where this line cuts the V.L. of the plane receiving shade is the vanishing point of the shadow—the V.S.*

The rule is really the same as that for the artificial light ; the use of such terms as V.P. and V.L. alter its outward appearance. For when we draw any line to a V.P. do we not draw a line parallel to the lines properly going thither, and if we draw this line from the V.P. of the sun's rays, have we not done precisely what our former rule bade us do? Again, our former rule told us to find the radiation points *on* the planes receiving shade. Now and again we had to go right to the far edge of the plane to obtain contact—that is merely what we do now when we make our line strike the V.L. of the plane.

Our work, similar as it is, is complicated by the vanishing⁊ and not vanishing of our various lines, and of the sun's rays.

53.

## To find the V.P. of the Sun's Rays.

The rays do not always have a V.P. When they are parallel to the picture, they do not. They are then said to be parallel to the plane of the picture, and at such and such an angle to the ground. When this is the case, the shadows are very easy indeed to obtain. Paragraphs 54 and 57 deal with them.

When not parallel to the picture, the rays are in a vertical plane at such and such an angle to the picture, and make such and such an angle with the ground.

Of course the sun could occur at the C.V., or so directly behind the spectator and so low that its rays vanished there. It may be above the C.V. at any angle to the ground. It is then in a vertical plane perpendicular to the picture.

Students who can obtain accidental vanishing points will have no difficulty whatever in finding the position of their V.P. of sun's rays.

When in a vertical plane perpendicular to the picture

the first method of Fig. 119 is used ; that is, we have no
direction to left or right to obtain.

When the sun is said to be, say, in a vertical plane
receding to the right at 40° and inclined to the ground
at 30°, then the second method of Fig. 119 is used.

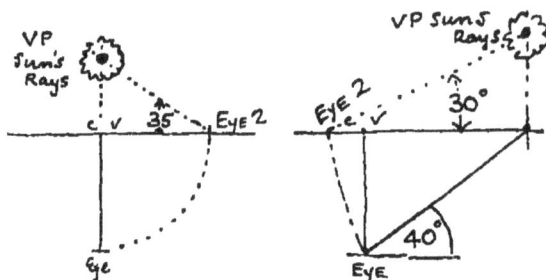

FIG. 119.—Finding V.P. sun's rays.

One thing only has to be remembered further, that
if the sun is *behind* the spectator (or *before the picture
plane*, which is the same thing), the V.P. of its rays are
found *below* the horizon and on the opposite side. In
that case we do not see the sun itself.

## 54.

## Shadows of Lines without V.P.'s when the Sun has no V.P.

When these conditions obtain, the form of the shadow
is governed entirely by the plane receiving shadow. The
shadows are drawn parallel to the V.L. of the plane.

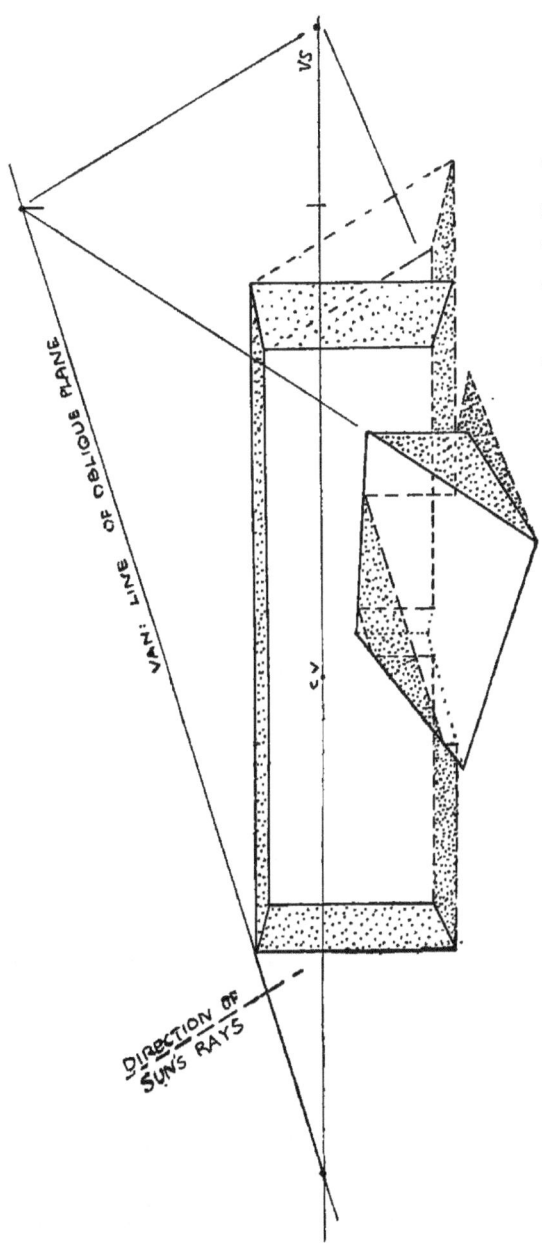

FIG. 120.—Shadows of line parallel to the plane of the picture, the sun being also in the plane of the picture.

In the diagram the shadow of the long horizontal plank is seen passing along the ground and over a prism, the upper surface of which is in an oblique plane, its V.L. being given.

The trail of the shadow across this prism is parallel to the V.L.

The sloping edge of the oblique plane casts a shadow vanishing to V.S., which is found by drawing a line from the V.P. of the edge parallel to the direction of the sun's rays.

## 55.

## Shadows of Lines parallel to the Picture with the Sun behind the Picture.

The object in this example is composed chiefly of lines parallel to the picture—horizontal, vertical, and inclined, and all of which are consequently without vanishing points. The object also has a few lines not parallel to the picture, but vertical to it. These lines all vanish to C.V., and the shadows of none happen to be seen.

The object or arch on its left side illustrates the shadows obtained on the ground plane of lines which do not vanish. From point A upwards is a vertical line; from the top of that proceeds an inclined line for a short distance, and from its end again passes the long horizontal

line of the lintel. All these three lines cast shadows on
the ground. The lintel again casts its shadow on the

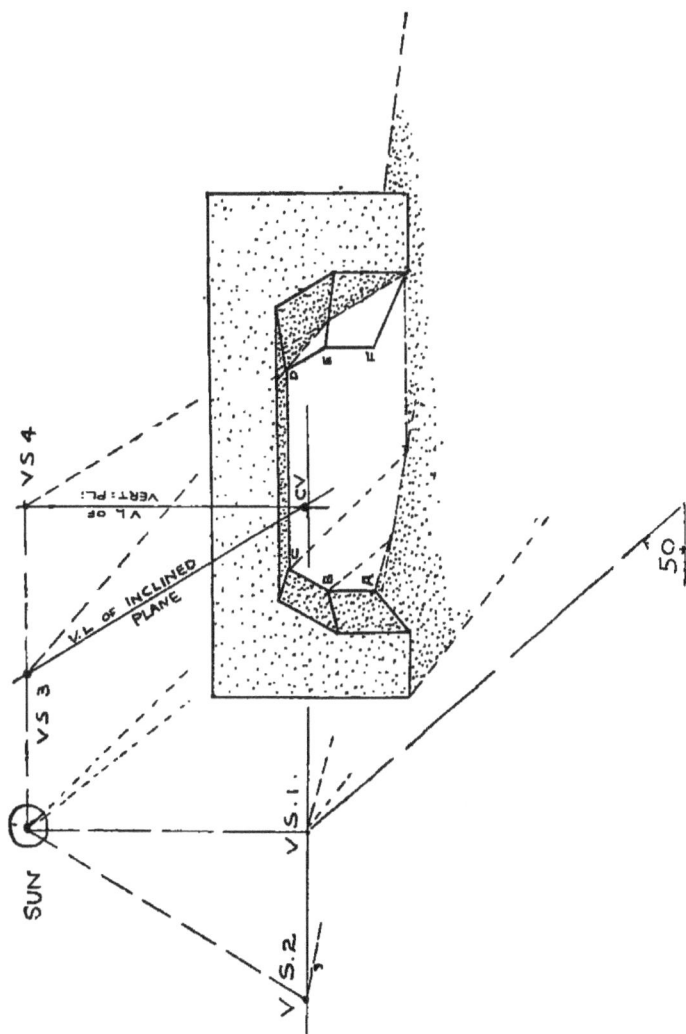

Fig. 121.—An object in parallel perspective with shadows cast from the sun, when sun is in a vertical plane making 50° with the picture toward left, and at an altitude of 30° to the ground.

vertical and inclined surfaces of the inner side of the
support on the right. The example includes, therefore,
several instances.

Returning to the shadows on the left side. If we apply our rule—

To find vanishing point of shadow (V.S.) of AB, draw through sun parallel to AB, *i.e.* vertically to the V.L. (horizon) of the plane of shadow (ground). We thus gain V.S.$_1$. The direction of the shadow of AB is given by a line from V.S. through A, and it is limited by a ray from the sun through B.

To find V.S. of BC, draw from sun at 60°, which is the angle of BC, to find V.S.$_2$.

A line parallel to CD, the lintel, will not strike the horizon anywhere, so the shadow of CD is drawn parallel to the horizon, and is not completed before the shadow is intercepted by the vertical surface of the right-hand end.

The shadow of CD on the surfaces DE and EF will be found in the same way. First draw the V.L.'s of these two planes. DE is in an inclined plane perpendicular to the picture, therefore we gain its V.L. by drawing parallel to it through C.V. The same process applied to the vertical surface EF yield its V.L. also. Having the two V.L.'s, we apply the rule. Draw from sun parallel to CD, and where it cuts the V.L.'s are the V.S.'s; V.S.$_3$ being V.S. of CD on DE, and V.S.$_4$ being V.S. of CD on EF. The shadow running down the surface EF will touch the ground where the shadow of CD along the ground was formerly interrupted.

## 56.

## Shadows of Lines which vanish and are cast by the Sun when its Rays vanish.

The vanishing points of the shadows are found in precisely the same way as formerly. The V.S. of a line occurs on the V.L. of the plane receiving the shadow, and is found as before by crossing the V.L. by a line drawn through the V.P. of the line and the V.P. of the sun's rays.

In the example given it will be seen that there are four planes. Of these three receive shadows—the oblique plane, which receives most of the shadows; the ground, which receives the shadow PO; and the side of the object, which receives the shadow BD.

Looking along these vanishing lines, the student will see that the V.L. of the oblique plane contains V.S.4, 1, 6, 5, and 3; the horizon contains V.S.7; and the vertical V.L. contains V.S.$_2$.

The sun is *behind* the spectator, or in front of the picture-plane. The rays are in planes making 60° with the picture-plane toward left, and are inclined to the ground at 43°. The whole of the working by which the V.P. of sun's rays is obtained is shown.

The student will notice that the object contains vertical lines, such as B and MP and AD; horizontal lines, as CK and inclined lines, CB and MK. The

M

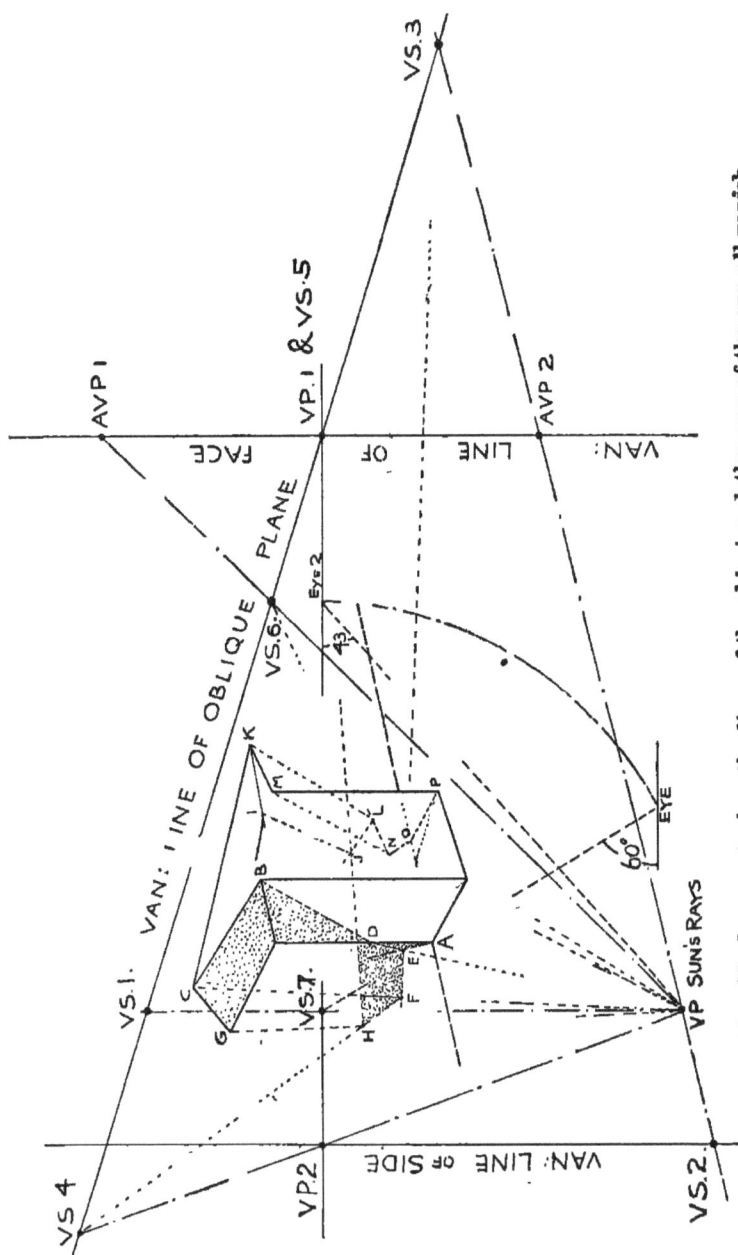

FIG. 122.—Shadows cast when the lines of the object and the rays of the sun all vanish.

oblique plane strikes the ground through the further edge of the base of the object.

The problem is worked as follows : Commencing with the verticals from the base, we find the V.S. A line drawn vertically through V.P. sun will strike the horizon in $V.S._7$ and the oblique plane in $V.S._1$. We see at once that a line from the nearest corner of the base to either of these will fall within the base, so we conclude that that vertical will not cast a shadow. Starting, then, at A, we vanish the shadow to $V.S._1$, because the oblique plane at once takes the shadow. Leaving this shadow for a moment, we get the shadow of BC on the side of the object. The V.P. of BC is $A.V.P._2$. We therefore draw through the V.P. sun, and it, to the V.L. of the side of the object, and so obtain $V.S._2$. We can then draw BD, and we find that the shadow is not complete before it has come to the edge D.

We transfer D by a ray to the V.P. sun, and this secures E. AE is then the shadow of AD. We return again to the shadow of BC, which now is on the oblique plane. We find a new V.S. for it in $V.S._3$. We limit the shadow at F by a ray from C.

Next we take CG. This vanishes at $V.P._2$, so we find $V.S._4$ on the V.L. of the oblique plane. The rest are obtained in exactly the same way.

## 57.

## Shadows of Lines vanishing, the Sun's Rays not vanishing.

The rule is again the same, and applied in the same way, if the rays do not vanish but the lines do. In the example, lines vanish to V.P.$_1$, V.P.$_2$, and V.P.$_3$. The shadow falls on an oblique plane, whose V.L. is given. The vanishing points of the shadows are found by drawing lines through V.P.$_1$, V.P.$_3$, and V.P.$_2$, to cut V.L. of shadow in V.S.$_1$, V.S.$_3$, and V.S.$_2$.

Point A on the oblique plane is given. The shadow is obtained by drawing the lines of the shadow to V.S.1, 2, or 3, and limiting these lines by rays drawn parallel to one another and in the direction required.

It would not matter at what angle the plane receiving shade were to the ground, the V.S. would always be found by continuing the lines which are now joining V.P. to V.S. till they cut the V.L. of the plane. In this way several V.S.'s, on as many V.L.'s, would be got by a single line from a V.P.

## 58

## Shadows of Circular Objects.

Circular objects, such as cylinders, cones, spheres, and niches, having no lines or edge, do not present the

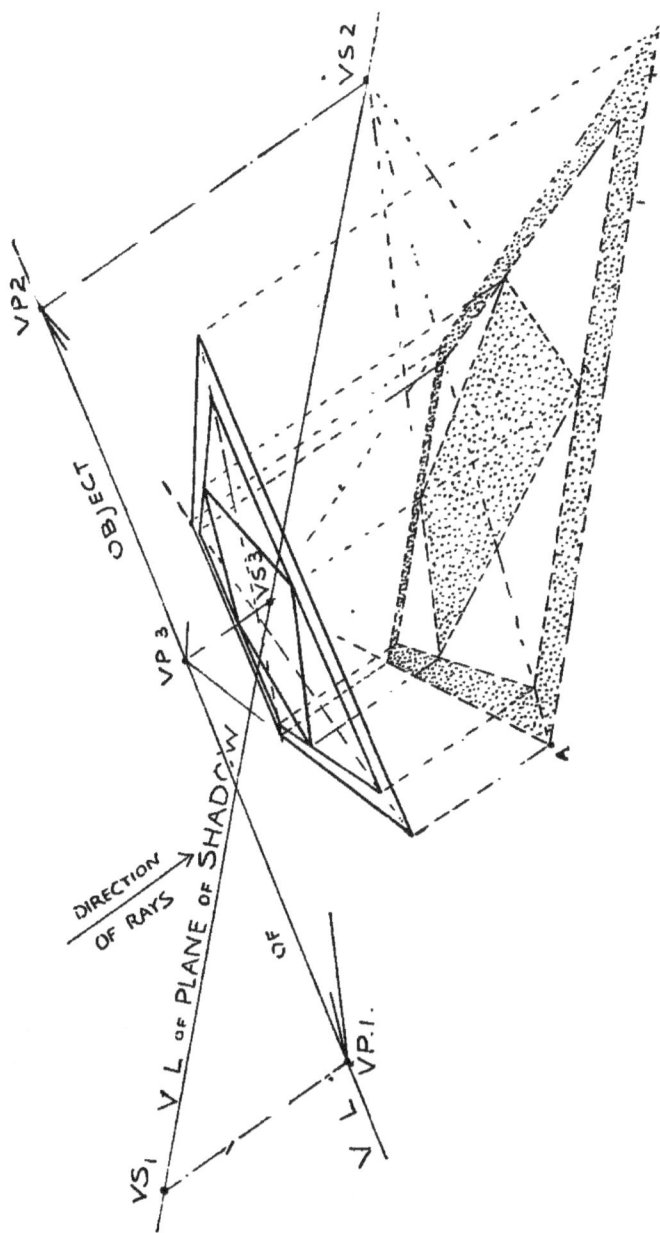

Fig. 123.—A square within a square, vanishing to V.P.'s 1, 2, and 3. The sun in the plane of the picture at the angle to the ground given. The shadow is to fall on an oblique plane.

features by which our rule can be applied.  In the case
of a cylinder, for instance, we have a V.P. of its axial
direction (V.P.$_1$ in Fig. 124), and we have a V.P. of the
planes of the ends at right angles to the axis.  But it
is obvious that V.P.$_2$ has really nothing to do with the
form of the cylinder, and any point over or under V.P.$_2$

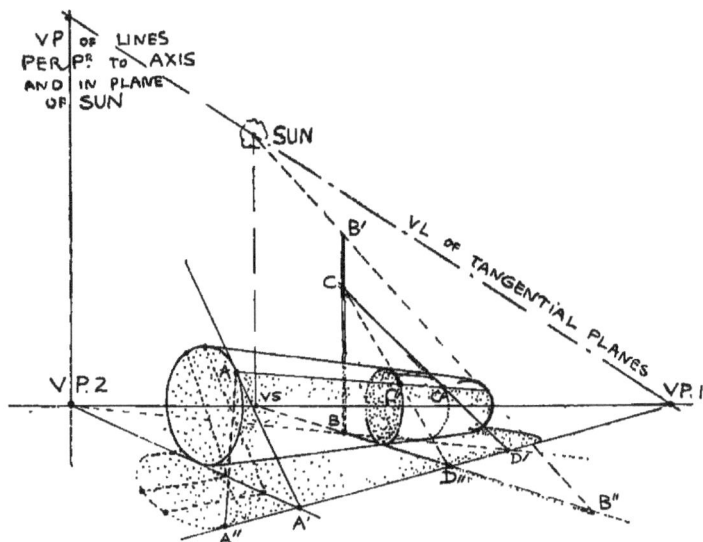

Fig. 124.—Shadow on and from a cylinder.

has as good a claim to represent the circular end.  V.P.$_2$
is, in fact, the V.P. of the intersection of the plane of
the end of the cylinder with the ground.  A shadow
can always be found by getting the shadows of verticals,
the summits of which are points in the object, so that
the shadow of the cylinder in Fig. 124 could be found
by taking points anywhere on the circumference of the
end, dropping verticals down to the ground, and thus
finding their seats on the ground.  From the verticals

thus found and posted on the ground, could be readily obtained the shadows they would cast.

' After all, this is what is done in Fig. 124, although the verticals are slanting. Thus A is assumed anywhere. AA' is a line touching A, and touching the ground at A'. Find its shadow by the rule. Draw through the V.P. sun a line parallel to AA', that is, from its V.P., and produce it to cut the ground at its distant edge, the horizon. V.S. is thus formed at V.P.$_1$. Then draw A'A'' as the shadow of AA', and find the limit A'' by a ray from the V.P. sun through A.

It is always advisable to employ such lines as accord with the object's form—lines too, which, taken in series, accord with the object's form. Now, if our line AA' be *not* in the same plane as the circular end, while it itself would serve us very well, other lines to give other points, if parallel to AA', will give a great deal of trouble. It would be very difficult to know where such a medley of lines touched the ground. But if AA' be assumed in the plane of the circular end, then it will strike the line from V.P.$_2$ through A'. And not only it, but lines parallel to it, will do the same.

For such reasons of convenience, we assume our line AA' as such that it (1) is in the plane of the end of the cylinder, and (2) is in the plane of the sun. The V.P. of AA' is readily found by drawing from V.P.$_1$ through V.P. sun till the vertical vanishing line over V.P.$_2$ is cut.

The usefulness of lines fulfilling these conditions is seen if we consider the shadow of the stick BB' cast by the sun across the cylinder. This vertical stick would have its shadow found by drawing from the seat of the sun through B to B″. B″ is found by drawing a ray from V.P. sun through B′.

Now, the shadow BB″ is not perpendicular to the axis of the cylinder—it cuts athwart it; and if we may speak of the shadow climbing over the cylinder, it does so in a sectional curve which is *not* a circle. This section is shaded in the diagram. To have to obtain an elliptical section for such a purpose as finding the trace of the shadow of the stick over the cylinder would be excessively troublesome. If, however, the section is obtained from V.P.$_2$, it will be circular and easily obtained. Obtained it must be, by careful working, which is omitted for clearness here.

If on this section we assume point C′, and run line C′C″ along the cylinder, using V.P.$_1$, we have a line which can be cut by rays from the sun. We then find what point on the stick corresponds to C′, the tangential limit of our shadow, by drawing a line from C′ to the V.P. of lines perpendicular to the axis.

It gives point C. Then by a ray from the sun we carry C to C″ on the cylinder. By taking other (any) points on the stick below C, finding by means of the V.P. of lines perpendicular to the axis the points on the section similar to C′, we repeat the same process, and obtain several points in the shadow of the stick and draw through them.

A cylinder standing vertically, as a column, would have the shadow of any projection above it cast as in Fig. 125.

The relation between the projecting edge upon which B occurs and the shaft is by a horizontal surface. Line BA is a horizontal line. The V.P. of horizontal lines in the plane of the sun is sought. It is the V.P. of the sun's direction. From this V.P. any lines are drawn from the top of the column to the projecting edge, as from A to B. From A a vertical is run down the column, and from B a ray is drawn to the V.P. of sun's rays. C is a point in the shadow.

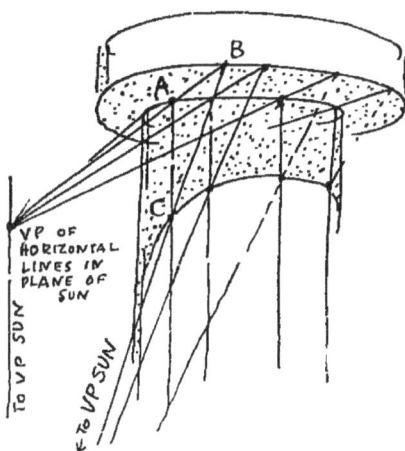

Fig. 125.—Shadow upon a column.

The shadows on, from, and on to a sphere have to be found by means of sections of the sphere. The perspective representation of a sphere is itself properly found by a series of sections. That is, a number of circles in vertical planes, and having their centres coinciding and being all of the same size, are put into perspective, and then the form of the sphere is drawn by freehand around them. The shape produced is not a true circle, and does not command much admiration.

The shadow on a sphere forms a circle just as a section of a sphere does. The circle separates the zone of light from the zone of shadow, and is drawn in perspective precisely in the same way as were the various circles by which the sphere's form was obtained if that process

has been adopted. This circular section will be perpendicular to the rays of light. In Fig. 126 the circle
is drawn, but the working is not shown. In that case
the V.P. of sun is down on the left, and planes perpendicular to the rays will be oblique planes with their
V.L. passing through the two V.P.'s of perpendiculars—
V.P. of perpendiculars to sun's rays, and V.P. of perpendiculars to direction of sun. The circle for the
shadow would be found in the same way as the chess-
board of Fig. 96, the centre of the sphere determining
the position.

Circles are not actually drawn in perspective, but
are obtained by putting into perspective a framework
of straight lines as in Fig. 150. In casting the shadow
of a circle, then, one neglects the circles and finds the
shadow of the lines as if they were the real object. The
object, thus being in straight lines, is worked as in
the preceding paragraphs—in fact, the working of
Fig. 123 is very much what one would use.

If no square framework were used, such a method as
that in Fig. 126 could be employed.

Having the circle which divides the zone of light from
the zone of darkness—that is, having the *shadow-line* of
the sphere, take any point A; draw AB and then BC to
V.P. of sun. C is on the line from V.P. of sun's direction
to X. From C draw CD, and from A to sun to find E—
a point in the shadow.

In Fig. 127 AB and AC are two edges of a square
which throws a shadow on the sphere. Either on the
ground draw a plan of the sphere, or through the centre

of it draw a horizontal section.  On the ground *ab* and
*ac* are the plan of AB and AC.  To find the shadow of

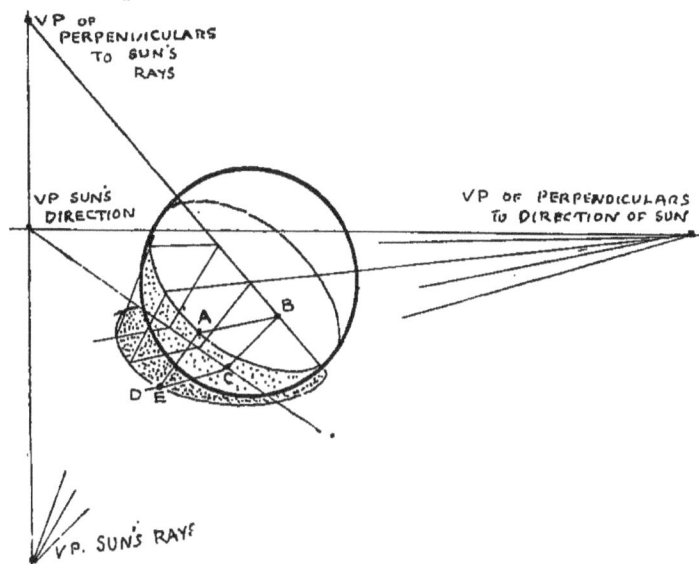

FIG 126.—The shadow from a sphere.

point A: drop A to the ground, to *a;* draw *a*K to V.P.
of sun's direction, not to V.P. of AC; raise KM, project MN

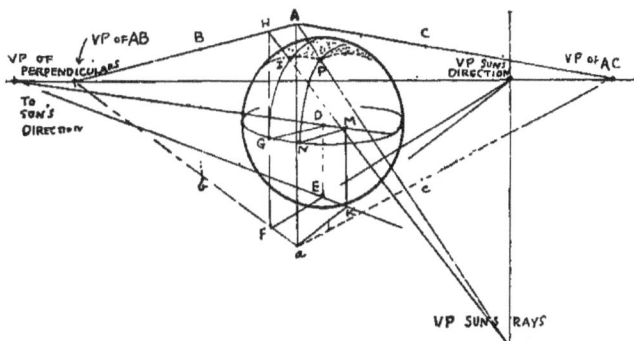

FIG. 127.—A shadow cast upon a sphere.

from V.P. of sun's direction, over NM draw a section of
the sphere in the plane of the sun's direction.  Then the

arc NP will be exactly between point A above and V.P. sun's rays below. Join A to V.P. sun's rays and P is a point in the shadow.

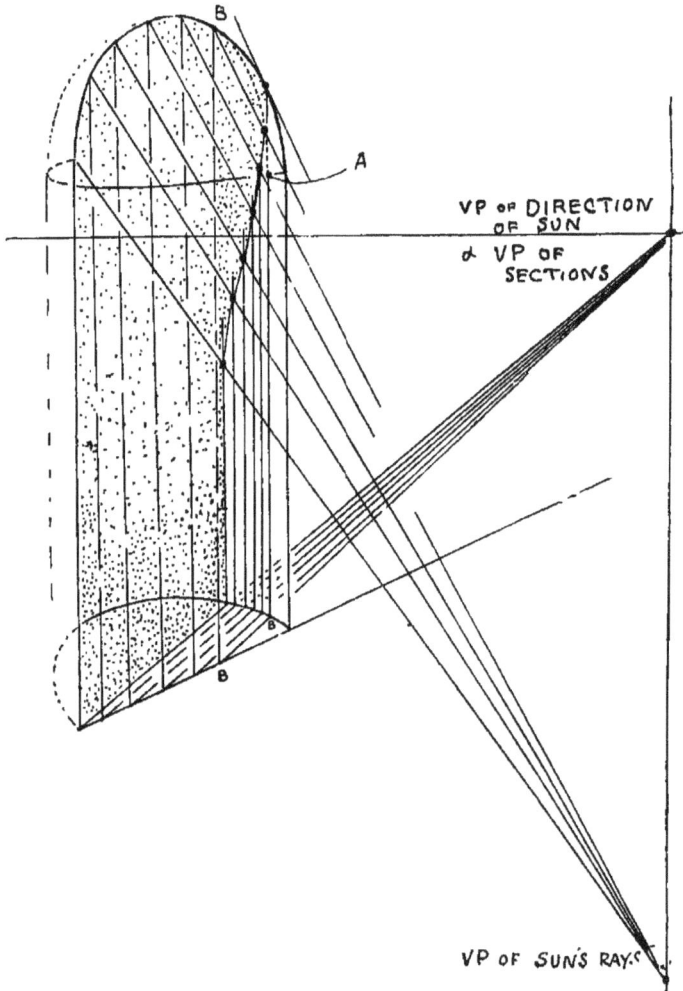

FIG. 128.—The shadow in a niche.

This is treated in the same way as the preceding examples. Various points are taken in the curve of the niche, and are then dropped vertically to the line on the ground. These vertical lines are the front edges of section-planes running back to V.P. of sun's direction. In their course they strike upon the inner surface of the niche. If the sections were all fully drawn they would run up vertically till the hemispherical head was reached, and then would continue within the hemisphere as curved lines. Only one of them, that for point B, needs to be so continued, and its curve commences at A. The other shadows fall on the vertical surface.

Shadows of and on such objects as these, when cast by artificial light, are found in precisely the same way. The only difference is that, instead of the vertical line from the sun up to (or down to) V.P. of sun's direction, we have a vertical line near at hand with the actual light at one end and the foot or seat of the light at the other. The sections will not be parallel, but will radiate from the seat of the light, but this makes no difference.

## 59.

## Reflections.

Reflections fortunately do not give much trouble. A reflection is an image, and an exact image, of the object reflected. It is the exactness of the similarity which makes the procedure simple, for the draughtsman has already all the points needed for the vanishing and

measuring of the original object, and these serve again;
or, if they do not serve, the new points are very easily
found from the old ones, and do not have to be found by
a laborious process.

The law upon which our methods of finding reflections
are based is this: *The angle of incidence is equal to the
angle of reflection.* So that angle ABC, Fig. 129, is
equal to angle DBE. If we make underneath the object
in Fig. 129 a similar object reversed, we have EF equal

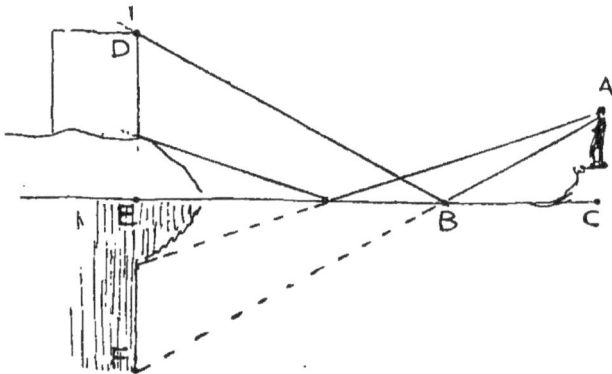

FIG. 129.—The theory of reflections.

to ED, and the triangle EFB equal to EDB; so that
angle FBE will equal angle DBE. In a word, ABF will
be in a straight line, and it will be as though the spectator saw a second image reversed immediately below the
first, the dividing-line being the mirror—in this case a
sheet of water.

Vertical lines are simply reversed. Thus the tree in
Fig. 130 has its foot on the level of the mirror (let us
suppose). Vertically down from the foot of the tree we
can put the height of the tree, and so get its reflection.

This is done throughout Fig. 130. Care must, of course, always be taken to reverse the measurement from the *level of the mirror* and not merely the foot of the object, unless the two coincide. Thus in Fig. 131 the edge of the building is not on the water-level. It has, therefore, to be dropped to that level, and the bank is represented

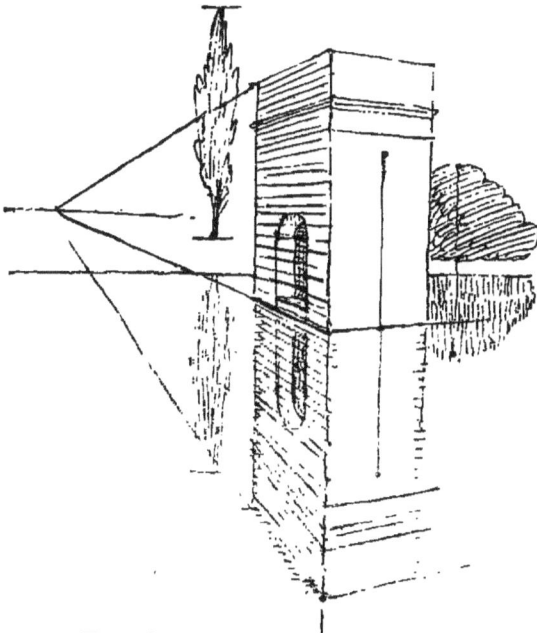

Fig. 130.—Vertical heights reflected.

as cut away to show the foot more clearly. In Fig. 132 the reflection of a bird is shown. This diagram shows the working in geometric form. The sloping roof vanishes to A.V.P. The V.P. of reflection is as far below the horizon as A.V.P. is above it.

The two most important results of the correct drawing of reflections are these—

1. Objects at different distances in reality are seen in a different grouping in the reflection than in the actual scene. Thus, Fig. 133, a figure lost amid a low-toned background comes out distinctly in the reflection, because the trees are too low and too distant from their reflection to " come in."

2. Objects of which the spectator sees the upper side

FIG. 131.—Reflection of a building above the water-level.

reveal the under side in reflection. The most usual example is a boat (Fig. 131), but the fact is seen also in the figure in Fig. 133, where above, the back of the hand is seen, below, the palm. The bent leg also undergoes considerable change; in short, the reflection shows the figure as if it were looked at from below.

Of reflections in vertical mirrors the simplest (and ugliest) are when the mirror is perpendicular to the picture and the object is in parallel perspective (Fig.

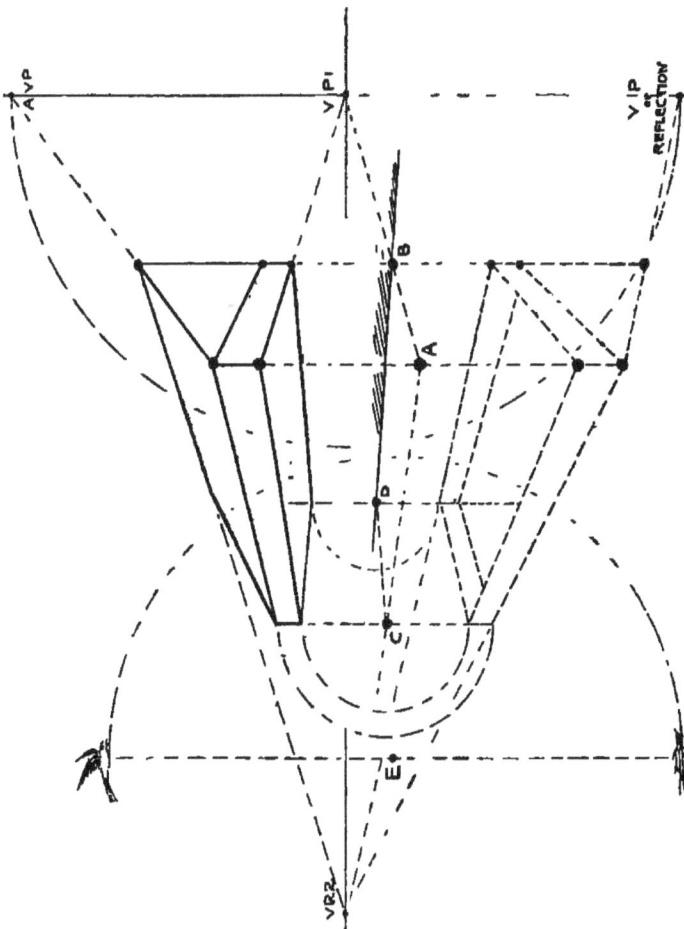

FIG. 132.—Reflections in a horizontal mirror, such as a sheet of water. A is a point on the surface of the water under the nearest corner of the object. E is a point on the surface of the water under the bird.

134). There the object vanishes to C.V., and the image will also, because, *and only because*, the mirror vanishes there too. We see, there, the image level with the object, the measurements equal, and taken equally from where the lines of the object, projecting toward the image, touch the mirror.

Fig. 133.—The reflection different in grouping and effect from the object itself.

Whenever the mirror vanishes to C.V., the projecting lines from object to image are parallel with the horizon. This is seen also in Fig. 135—there V.R. is the same distance to the right of C.V. as V.P. is to the left of it.

If, however, the mirror is at an angle to the picture as in Fig. 136, then two things happen—

1. The projectors BC and CD will vanish to V.P. of perpendiculars to mirror. This is a V.P. found in the same way as the V.P. of perpendiculars to vertical planes inclined to the picture.

2. The reversal of the V.P.'s must be effected at the eye. Thus, in Fig. 136, B is a line vanishing to A.V.P., which is over V.P.$_2$, its V.P. of direction. The vanishing parallel from V.P.$_2$ to eye happens, quite accidentally, to pass through B, so can readily be identified. It makes a certain angle at the eye— an angle indicated in black—with the vanishing parallel of the mirror. This angle, being repeated on the other side of the mirror, and also

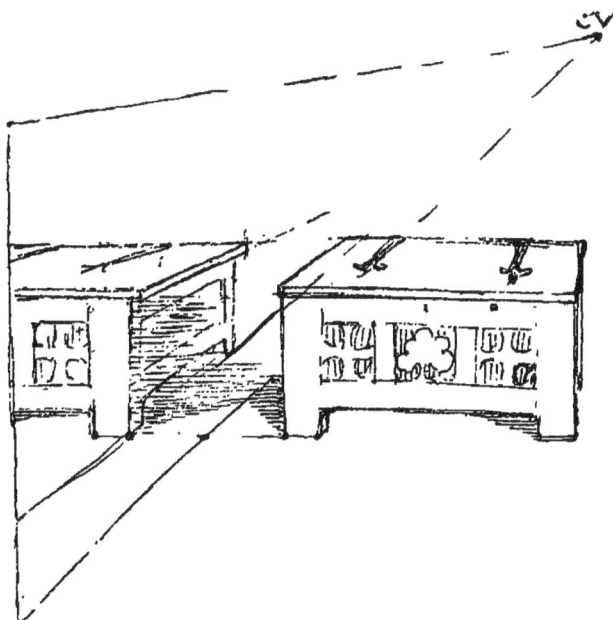

Fig. 134.—The mirror vertical and perpendicular to the picture. The object parallel to the picture.

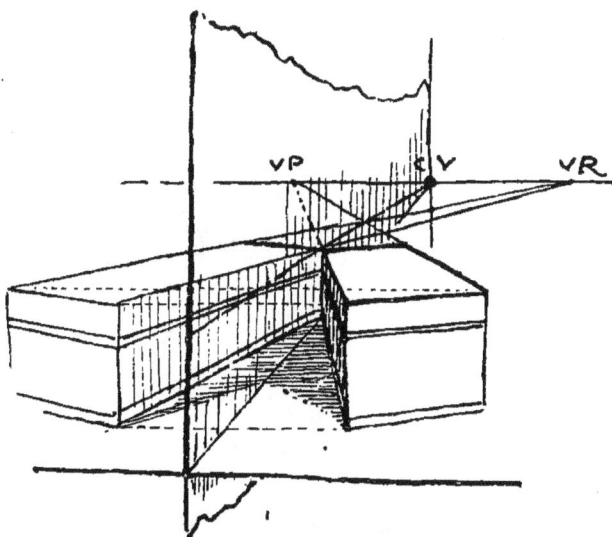

Fig. 135.—The mirror vertical and perpendicular to the picture. The object in angular perspective.

blackened, gains us V.P.₃. To get the inclination of the
V.P. of reflection over V.P.₃, a new eye is required. It

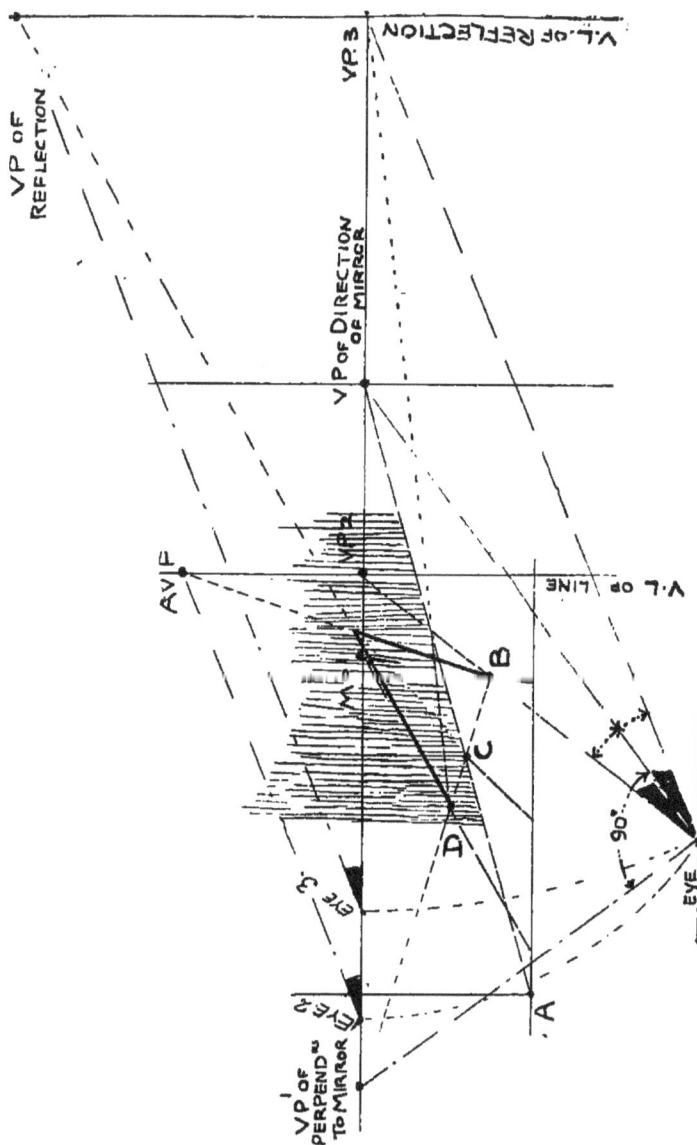

Fig. 186.—The mirror at an angle to the picture.

is eye 3, and the same angle is set up at it as was set up at eye 2.

The perpendicular CD has to be measured by a measuring point, M.P.₁.

## 60.

## Aёrial Perspective.

The only fact of aёrial perspective which the student need trouble about—and it is of comparatively little value—is that objects of the same strength of tone or darkness become paler as they recede into the distance. This paling is due to the interposition of the atmosphere. The mistier the day, the more sudden the vanishing of the tone.

## 61.

## Solving Problems.

The problems, the solutions of which are given in the following pages, are the problems contained in the two Examinations of the Board of Education in 1901. The questions are so admirable that it were foolish to seek others, particularly as one can at the same time place before students the difficulties which candidates have had to face. The questions, moreover, introduce a large variety of subjects, many of which would have to be treated specially if they were not included in these examples.

The Syllabus of these examinations is printed on pp. xiii. and xiv. After reading the Syllabus, one is not surprised to find the paper in two divisions, A and B, A for working, B for sketching, and the following General Instructions prefacing the questions themselves :—

GENERAL INSTRUCTIONS.

**Before commencing your work, you must carefully read the following instructions :—**

*You may not attempt more than five questions, of which one must be, and two may be, taken from Section (A).*

*All your drawings must be made on the single sheet of drawing-paper supplied, for no second sheet will be allowed. You may use both sides of the paper.*

*Shadows and reflections may be indicated in outline only,*

*or washed in with a light tint. Pencil shading is not to be resorted to. None of the drawings need be inked in.*

*Put the number of the question close to your workings of problems, in large distinct figures.*

*The number of marks assigned to each question is stated in brackets.*

*Your name may be written only upon the blue slip attached to your drawing-paper ; it is not given to the Examiner, and you are forbidden to write to him about your answers.*

After these General Instructions follows a brief preface to Section (A) as follows :—

## SECTION (A).

*Read the General Instructions.*

*You may attempt two questions only, which must be solved accurately, using instruments.*

*All lines used in construction must be clearly shown.*

The problems which follow are most of them accompanied by diagrams, either as a geometric method of describing the object to be represented, or being a portion of a perspective drawing which has to be completed.

In the following pages no attempt at working the problems is made. It is better that a student gets his own results, and gets to know they are right by understanding what he is doing. Worked problems often are merely copied by students, not studied. It is hoped, therefore, that the student will not feel the need of what the author believes to be rather a hindrance than a help —the mere solution.

Here, once more, it may be pointed out how valuable it is to sketch a problem, even such as is usually worked. To do the problem by freehand will assist the student sometimes much more than by accurately working it, because the absence of the clear lines makes the interest centre in the why and how rather than in the tidy and plausible look of the drawing.

Two hours are allowed for working the paper.

### QUESTION 1 (April, 1901).

*Diagram Q. 1 shows in perspective a tower, square on plan, standing on the ground-plane, its nearest edge marked AB, and, raised above the ground, a horizontal plane on which is a point $A^1$. The horizon is at H.L. and C.V. ; E is the distance of the eye from the picture. At point $A^1$ erect a vertical line perspectively equal to AB. This is to be the nearest edge of another similar tower standing on the raised plane and having the same dimensions, and its sides inclined to the picture at the same angles as the tower shown.*

(10 marks.)

*The horizon must be drawn across the short way of the paper, 8 inches from the top, and the centre of vision placed $4\frac{1}{2}$ inches from the right-hand edge of the paper.*

We must look in all cases to the completion of the scaffolding of the problem ; that is to say, we must find horizon, eye, ground or picture line, two V.P.'s, and their two M.P.'s.

The horizon is given.

The eye is given.

The ground-line we draw through point A horizontally.

The distance of eye before the C.V. is given, and it will remain that length whether we place the ground-line representing the picture-plane through A or through

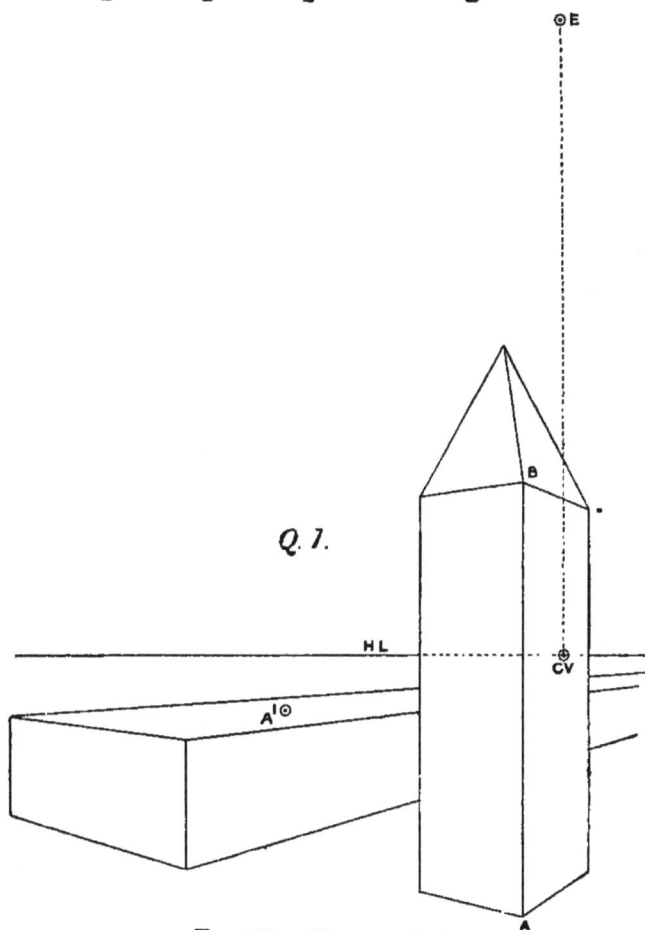

FIG. 137.—Diagram Q. 1.

the nearest corner of the raised plane beyond. If the ground-line were through the lowest (nearest) corner of the raised plane, its distance below the H.L. would be the same in fact, but less in the drawing. Hence the length *in feet* of the P.V.R. (C.V. to eye) will be able to vary. Compare Fig. 55, p. 81.

No scale is stated. We are, therefore, quite free, and draw the G.L. through A.

The two V.P.'s we find by continuing the lines to right and left till they meet on the horizon. It will be best to see that the vanishing parallels from these V.P.'s to the eye enclose a right angle, as they should.

The problem requires us to construct at $A^1$ a tower precisely similar to the one shown above A. It is to be the same in its dimensions, yet, being evidently further away, we must expect diminution of size on the drawing.

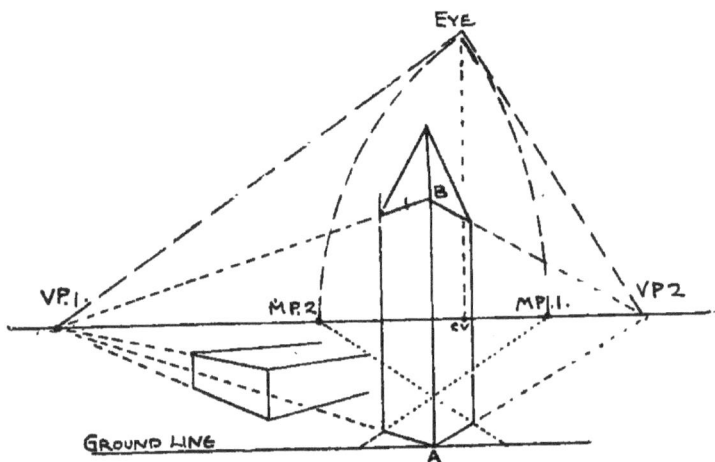

FIG. 138.

Then it is to stand on a higher level, so we must expect it to reach higher up the paper. Lastly, its sides are to be inclined to the picture at the same angles as the tower already given. Its horizontal lines will therefore vanish to the same V.P.'s as do those of the former tower, and we must expect to see the side which vanishes to the right more fully than the side on the left.

Before we can proceed we must find what the dimensions are of the tower already drawn. We require the M.P.'s.

We find the M.P.'s by placing the point of the compass on V.P. and stretching the pencil to eye and describing arcs to cut the horizon. Having found our two M.P.'s, we use them to obtain the true length of the short sides receding from A.

We now want the height of the apex of the pyramid as the top. Here we meet a "catch." The student will remember that heights are found by means of "lines of height;" or "walls" which run down to the picture, and are striped with the measurement required.

Obviously the two upright sides of our tower give us two such walls. One comes down from V.P.$_1$, the other from V.P.$_2$. Both these walls strike the P.P. in the line AB. Now, the student who is going to fail will do what is done in Fig. 139. He will use his V.P. quite properly, and bring a line from it through the

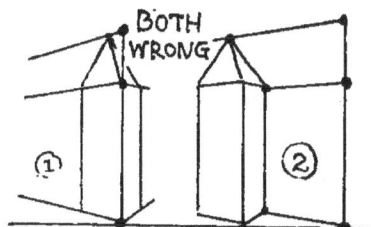

Fig. 139.

apex, but he will let it strike the line AB, lengthening AB for the purpose. The mistake lies in using the wall of the side for a point which is not in the side. The apex is some distance behind the "wall." Or he may do it as in No. 2 of the same figure; that is, use a wall at the back of the object. This is quite right as regards the square

part of the tower, but quite wrong as regards the apex
of the pyramid.

The wall of height must pass through the apex, through,
in fact, the axis of the whole tower. In Fig. 140 a plane
is shown passing through the axis and striking the
picture-line somewhere on the right of A. To find this
plane we merely find the middle of the side of the base

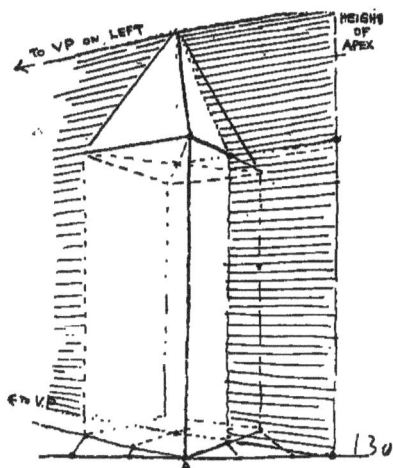

FIG. 140.—Finding the height
of the apex.

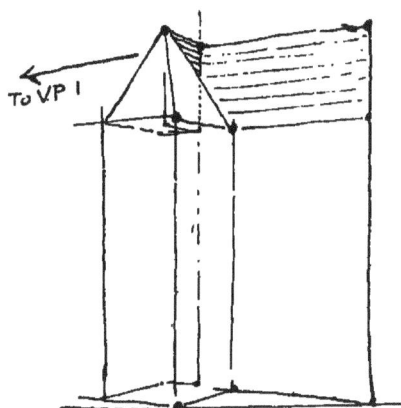

FIG. 141.—Using two walls of
height.

on the right, and bring a line down from the V.P. on the
left. This line will be under the apex, and therefore a
line through the apex will record the proper height on
the vertical set up for the purpose.

Or, two walls of height may be used, as in Fig. 141,
where a wall to V.P.$_2$ carries the height of the apex to the
vertical face of the tower, and then a wall to V.P.$_1$ brings
both the height of the apex and the height of the vertical
part of the tower forward to a line of height. The angle

between the two walls is over the centre of one of the sides in the rear.

One of the best ways is to use the V.P. of diagonals. This is done in Fig. 142. As a result AB lengthened upwards becomes the line of height, and line along the ground from the V.P. of diagonals carries in its course both two of the angles of the tower and its axis with the apex above.

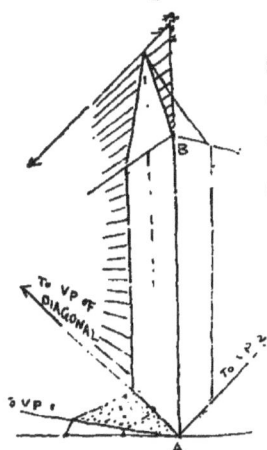

When we proceed to erect our second tower we meet another "catch." Point A', at which our tower has to commence, is raised above the ground, and we have to be careful how we measure the edges

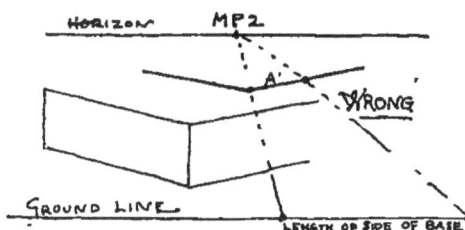

Fig. 142. — A wall of height vanishing to the diagonal V.P.

Fig. 143.—How *not* to do it.

and height of the tower. If we put in the base first, we proceed by drawing two front edges vanishing to V.P.₁ and V.P.₂ as before. The student who fails attempts to measure these lines by lines from the measuring-point down to the ground-line, as is done in Fig. 143, which is quite wrong. A' is not on the ground, so a line from MP₂ through it will not touch the ground-line.

Two methods are open to us. One is to make a new

ground-line, called a picture-line, which will serve for the raised plane exactly as the ground-line serves for the ground.

This method is shown in Fig. 144. If we knew the height from the ground of the raised plane, we could merely draw a new ground-line above the proper ground-line at the true height. In the absence of this knowledge

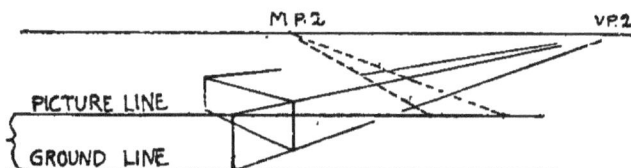

FIG. 144.—The picture-line, or new ground-line.

we proceed as follows. We continue one of the edges, on the ground, of the raised plane, from say V.P.₂ till it cuts the ground-line. We there erect a perpendicular which we cut by bringing down the upper edge of the raised plane, also from V.P.₂. This extends the plane right up to the front of the picture, and so gives us the proper height of it. Through the point thus found we

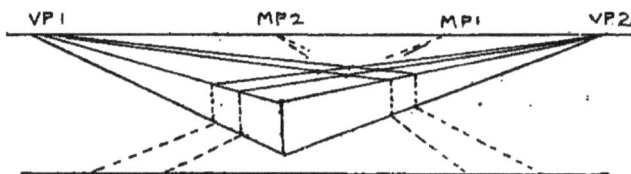

FIG. 145.—Lifting detail dimensions up from the ground-plane.

draw our picture line horizontally, and upon it we can measure the base of our new tower.

The other method is one whereby we trickle the sizes down and along the surfaces, and is shown in Fig. 145.

What is done there is this,—the point A' is transferred to the edges of the raised plane by bringing lines down through it from V.P.₁ and V.P.₂. From the upper edges it is dropped down to the lower, and then, along these lower edges, the sizes of the sides of the base of the tower are measured by means of the M.P.'s 1 and 2. The base of our tower happens to be parallel to the edges of the raised plane, and we are thus enabled to utilize the edges in this manner. The student must note particularly that the lines carrying the sizes on the lower plane are dotted lines going to the M.P.'s, while the lines carrying the sizes in the upper plane are lines going to the V.P.'s.

In obtaining the height of the tower care must be had to allow for the height of the raised plane. If a picture-line has been used as in Fig. 144, the bare height of the tower will be put up from it. If the "trickling" or "stair-carpet" method of Fig. 145 has been used, the height of the plane must be added to the height of the tower.

QUESTION 2 (April, 1901).

*Put into perspective on the ground-plane a rectangle 12 feet long and 8 feet wide, its longer side receding to the right at 40° to the picture-plane, the nearest corner being 3 feet to the left and 3 feet within the picture. This is the upper edge of a tank with vertical sides ; the bottom of the tank is a horizontal plane 1 foot below the level of the ground. In the centre of this lower plane erect a tapering shaft 12 feet high, its base 4 feet square, its top 3 feet square and parallel to the base, the edges of base and top being parallel to the horizontal edges of the tank. The top of the shaft is the base of a pyramid 2 feet high. The eye is 12 feet from the picture and 7 feet above the ground. Scale, ½ an inch to a foot.* (18 marks.)

*The horizon must be drawn across the short way of the paper, 5 inches from the top, and the centre of vision placed 7 inches from the left-hand edge of the paper.*

In all cases it is well to make a sketch before commencing work of the probable appearance of the object,

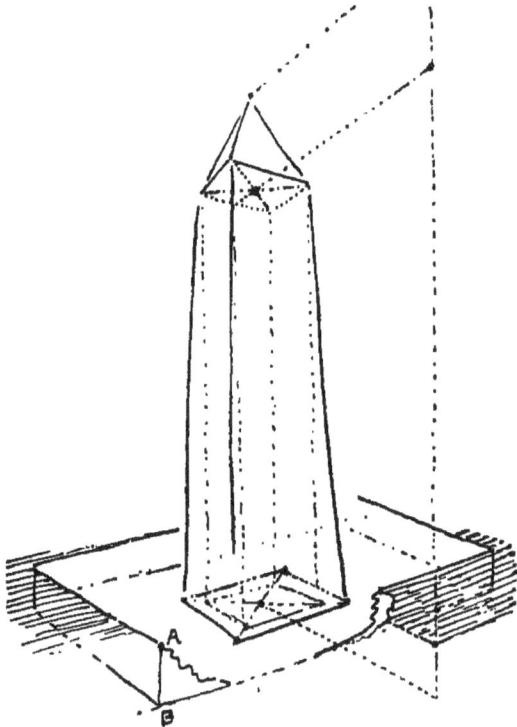

FIG. 146.—The subject of Question 2.

and especially should this be done when no diagrams accompany the question.

The subject of Question 2 is somewhat that shown in

Fig. 146. It is better, in making these sketches, not to attempt delicacies of foreshortening, but rather to make such drawings as reveal all the parts of the subject.

The specification requires us to first think of a rectangle as the ground-plane, and then to imagine another plane 1 foot below it, on which is the base of a tapering shaft. In practice it will be more convenient to at once go to the lower level and treat the bottom of the tank as our basal plane.

In setting out our scaffold for the problem, we may put our ground-line 7 feet below the edge, but it would be better to leave the ground-line out for the present, and put a new ground-line 8 feet below. This is done in Fig. 147, which is a sketch of the working. Upon this plane

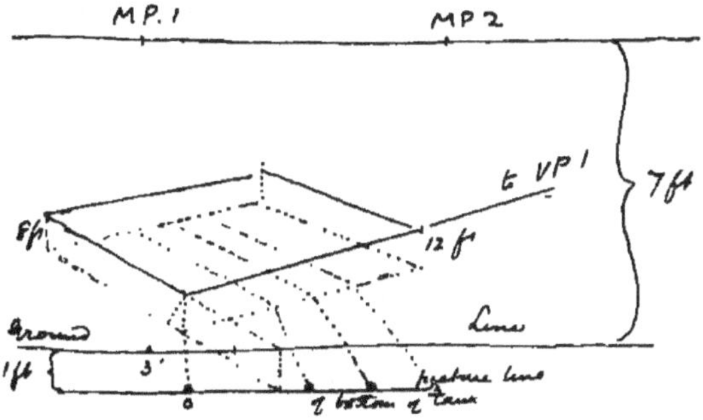

FIG. 147.

we proceed to get the rectangle with the square within it for the base of the shaft. We can at the same time, if we like, get the smaller square (3 feet) for the top of the shaft also on this low level. We could then raise each corner up to the requisite height. This would, indeed, be the best method of proceeding if the top of the shaft had to be so near the level of the horizon as to render

O

the working difficult. But in the present case we need
not do this unless we like. We can, that is, work the
top of the shaft from its own picture-line, as is suggested
in Fig. 148. In that diagram the centre of the top is

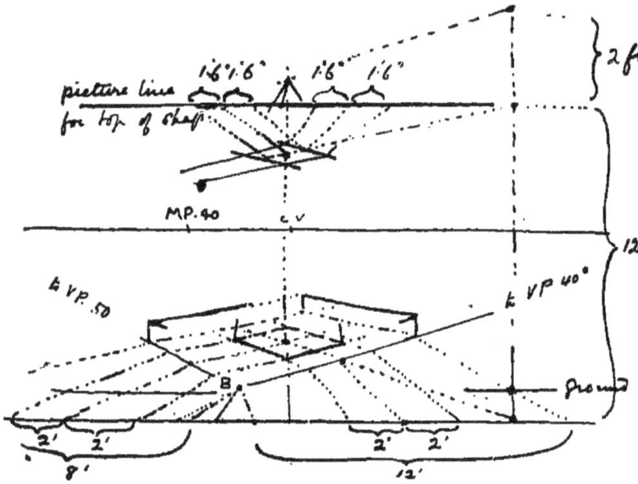

FIG. 148.

obtained by a vertical from the centre of the base,
measured by a line of height. This vertical serves also
to obtain the apex of the pyramid upon the shaft.

QUESTION 3 (April, 1901).

*Diagram ·Q. 3 gives the plan and elevation of a
circular disc pierced by a square prism. Put this into
perspective, the disc to be standing on the ground-plane,
the point of contact A being 1 foot to the left and 4 feet
within the picture. The face of the disc is to lie in a
vertical plane inclined 60° with the picture to the left.
The eye is 12 feet from the picture and 5 feet above the
ground. Scale, ½ an inch to a foot.* (20 marks.)

*The horizon must be drawn across the short way of the paper, 9 inches from the top, and the centre of vision placed 4 inches from the left-hand edge of the paper.*

The question does not give the sizes in figures of the objects, but leaves them to be found from the plan and elevation. Using the scale of ½ inch to 1 foot, the disc is 8 feet in diameter and 2 feet in thickness, while the square prism is 10 feet long. The prism consequently projects 4 feet before and behind the disc. The chain-lines on the diagram are not given to the candidate. They have been added because a circular object can only be treated by converting it into a polygon. The figures 0, 1, 2, 3

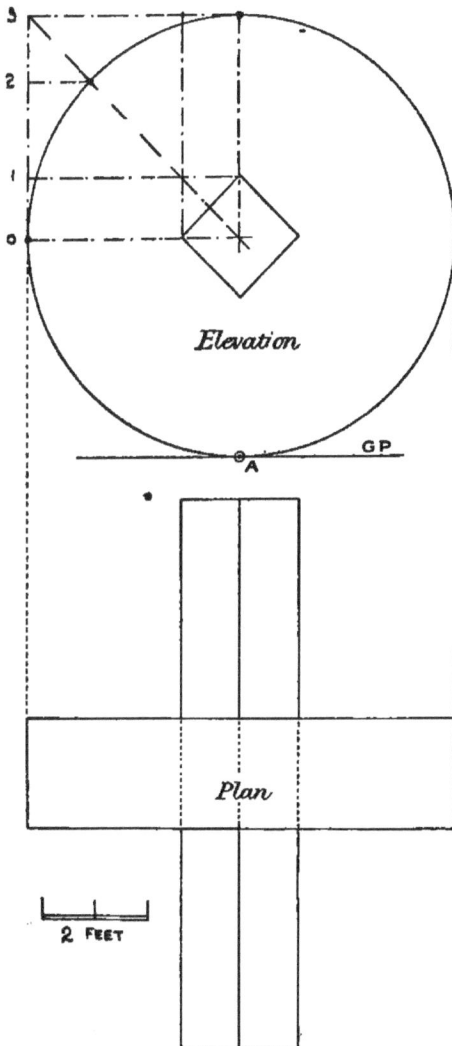

FIG. 149.—Diagram Q. 3. The chain-lines and figures are added.

are also added; as is the scale of 2 feet, originally ½ an inch each, but which in this publication should measure ¼ inch each. The scale of the diagram here given is about half the original scale.

If the problem is worked by horizontal lines only, the procedure will be as follows :—

Point A being found (by a line to C.V. from a point 1 foot on left, measured by the D.P.), a line for the base of the square is run down from V.P. 60 to the

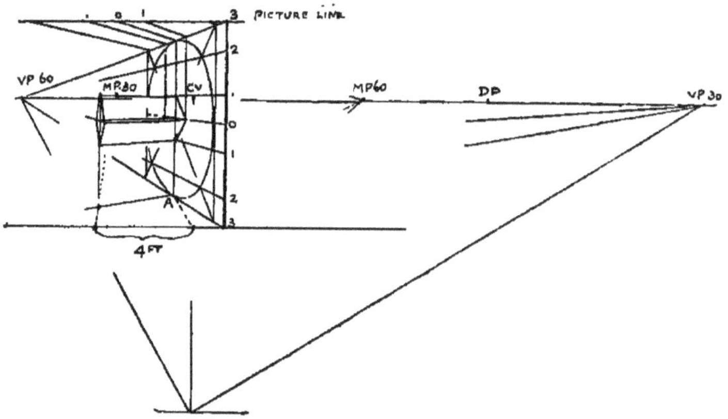

FIG. 150.—The problem worked by horizontal perspective.

ground-line. There a line of height is run up the height of the circle, 8 feet. It may be convenient to get a picture-line from the top edge of the square. This is done in Fig. 150.

The wall of height is then striped 3, 2, 1, 0, 1, 2, 3, as on the elevation. Next the vertical lines limiting the square are found. Point A yields the centre all up, and then by means of M.P. 60 the measurements are put off on the upper picture-line. The first line is that through the centre, O, and the others are put on either side of it. This must be the procedure, because the circle does not touch the line of height.

The diagonals being drawn, it is easy to find the points and draw the circle, which is done by freehand.

The prism is now projected forward, using V.P. 30. The 4-foot projection is first found on the ground by a line vanishing at V.P. 30 and measured by M.P. 30. This dimension is raised by a vertical which yields the upper, lower, and central points. Through the centre a line from V.P. 60 carries the side points out, and they are limited and the side edges drawn by lines from V.P. 30 through the corners of the square in the middle of the circle.

The other circle of the disc is found in the same way. A line from point A to V.P. 30 is drawn, and along it 2 feet are measured toward the right, just as 4 feet were measured toward the left for the square prism. We thus obtain a new point of contact, and proceed as before, being able, however, to transfer by means of the vanishing points many of the points we require, from the circle first obtained.

If, however, inclined perspective be employed, considerable saving is achieved—a saving more particularly of the pricking off of measurements.

The diagonals of the square for our circle and the sides of the ends of the prism are all parallel, and vanish upward or downward at 45° to the horizontal. *These accidental vanishing points* occur immediately over and under V.P. 60. They are found by first drawing

a vertical V.L. (for these lines are all in vertical planes)
through V.P. 60. Next the position of the eye in
relation to this V.L. is found; it coincides with M.P. 60.
Upward and downward then at M.P. 60 are drawn two
vanishing parallels, making 45° with the middle line,
the horizon. A.V.P.$_1$ and A.V.P.$_2$ are then the vanishing
points required.

The saving will be very apparent if the problem is

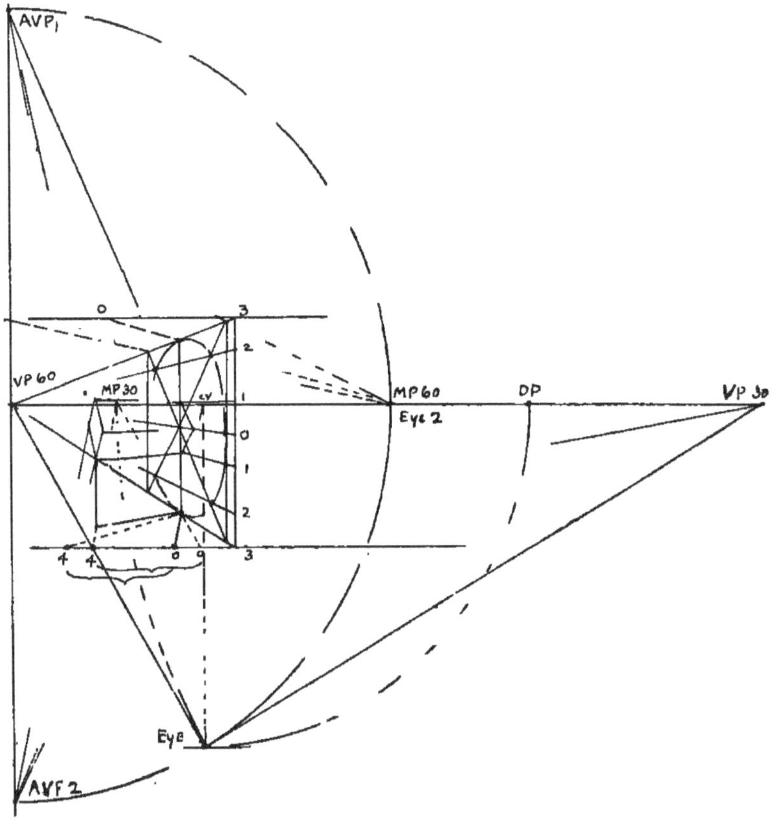

Fig. 151.—The problem worked by inclined perspective.

worked. If we get first the long line through A, and

which contains the base of our square, the long line
at the top, which is 8 feet above the ground, and a
line through the centre, all three vanishing to V.P. 60,
we shall find our A.V.P.'s help us to get the square.
Measurements are always easier to get by lines of height
than by lines to the measuring points. Moreover, they
are more reliable, for the measuring lines often pass
the lines they measure at very acute angles, and it
is difficult sometimes to exactly locate the point of
crossing.

There is no need to enlarge upon the utility of these
A.V.P.'s. If the student works the problem, and uses
them as freely as he can, he will find he can dispense
with many of the measuring lines otherwise necessary.

### QUESTION 4 (April, 1901).

*Diagram Q. 4 shows in perspective the main lines of
a barge (rectangular in plan and cross-section) lying on an*

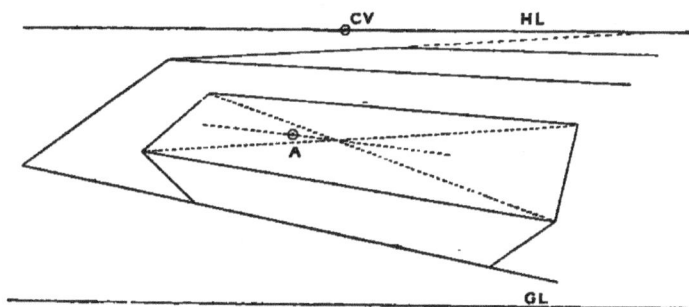

FIG. 152.—Diagram Q. 4. (Scale ¼-inch to 1 foot.)

*oblique plane, with one edge on the ground-plane, and a
point A on its upper surface. The position of the horizon,*

*the centre of vision, and the ground-line are given. The
distance of the eye from the picture is 11 feet. Scale, ½ an
inch to a foot. At point A erect a mast 8 feet long,
perpendicular to the upper surface of the barge.*

(27 marks.)

*The horizon must be drawn across the short way of the
paper, 7 inches from the top, and the centre of vision placed
4½ inches from the right-hand edge of the paper.*

This problem requires a knowledge of the treatment
of oblique planes, and involves in addition the finding of
perpendiculars to oblique planes, and the finding of new
picture-lines.

The first thing to do is to find the V.L. of the oblique
plane and all the V.P.'s. The student will remember
that an oblique plane is obtained by means of two V.P.'s
—one being the V.P. of the intersection of the plane with
the ground-plane, and the other the V.P. of lines running
directly up the plane at right angles to the intersection.
In other words, the second V.P. is the V.P. of the plane's
inclination to the horizontal plane, while the intersection
denotes the inclination to the vertical or picture plane.

Of course some indication must be and is given of the
positions of these points. The first indication is in-
corporated in the fact that the barge has one edge on the
ground-plane. This edge is, therefore, on the intersection
between the oblique and ground planes, for the barge is
lying (lying flat, that is) on the oblique plane. If, then,
we continue this lowest edge of the barge to the left till
it cut the H.L., we have the V.P. of the intersection.

Now, the V.P. of intersection is one V.P. of a rectangle; the other will be on the right. We have the position of the C.V. given; we have also the distance of the eye, 11 feet. We can thus find the position of the eye, and can set off toward the right a vanishing parallel at right angles to the one we may draw from eye to V.P.₁ already found. This parallel will give us on the H.L., V.P.₂, the V.P. of a direction at right angles to the intersection. The V.P. of inclination will occur somewhere over V.P.₂. We therefore draw a vertical line through V.P.₂.

An indication is given how far up the V.P. of inclination will be. It is given in the short ends of the

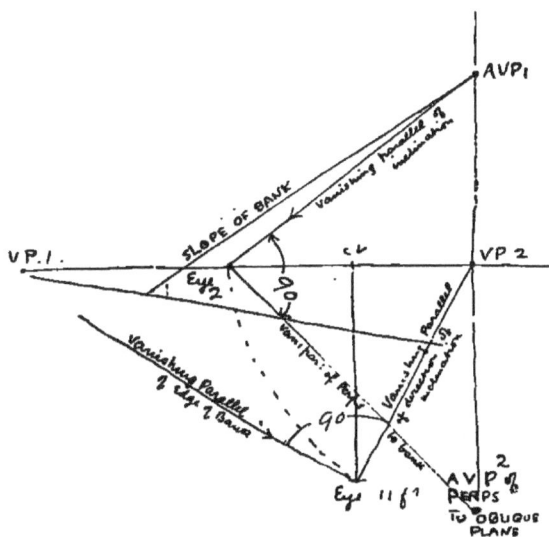

FIG. 153.—The V.P.'s found.

barge. We therefore converge these upon the vertical line, and so obtain A.V.P.₁. Through V.P.₁ and A.V.P.₁ we draw the V.L. of the oblique plane.

If the student will turn to paragraph 43, he will see how perpendiculars to oblique planes are obtained. If an oblique plane ascends, the perpendiculars to it naturally descend. If the plane ascends at 30° to the horizontal, the perpendiculars descend at 60°.

Perpendiculars to a plane are not treated as part of the plane. Excepting the point of contact, they are not part of the plane at all; they only bear a certain relation to it. We therefore assume the perpendiculars to be in some plane or other. Just as an upright line can be in any upright plane, no matter at what angle to the picture, so can a perpendicular to a plane be in a large number of planes. In Fig. 102, the student will see that there are three vanishing lines bounding a cube on an oblique plane. Two of these V.L.'s meet below at the V.P. of perpendiculars, and the V.L. of the planes to which these perpendiculars belong is the uppermost or third V.L. of the three. Using this V.P. of perpendiculars as a pivot, we might swing, or imagine swung, one of these descending V.L.'s across to the other. In its course it would cross C.V.L., C.V., and the eye in relation to the oblique plane all in one line. Perpendiculars to a plane can be in any plane whose V.L. contains their V.P.

Generally, when a choice has to be made where a direction is being assumed, the *vertical* is decided upon.

We already have a vertical V.L. through V.P.$_2$ and containing A.V.P.$_1$. What we want is an A.V.P.$_2$ at 90° to A.V.P.$_1$ above.

To find this angular measurement we require the

position of the eye in relation to this vertical V.L. The C.V.L. will be at V.P.$_2$, because that point is the nearest point on the V.L. to C.V. The distance of the eye from the C.V.L. will be that already existing in the vanishing parallel from eye to V.P.$_2$. In other words, the eye in relation to this V.L. will be at the M.P. for V.P.$_2$.

Deciding thus that M.P.$_2$ is the eye in relation to our V.L., we draw a line from it to A.V.P.$_1$. This is the vanishing parallel of A.V.P.$_1$, and shows the inclination of that point to the horizontal. We therefore set downwards from our new eye another vanishing parallel at 90° to this, and so obtain A.V.P.$_2$—the V.P. of perpendiculars to the oblique plane.

A.V.P.$_2$ being, then, the V.P. required, we draw a line from it through A on our barge, and this represents the mast.

The mast has now to be limited to 8 feet in length.

We want, therefore, the picture-line for the vertical plane containing the mast, and for which the V.L. has recently been found.

What picture-line have we already? Only one—the ground-line. If we can make the V.L. of our vertical plane cut the V.L. of the ground, we shall have a point *in both planes.* V.P.$_2$ gives us this point. Shall we then draw a line from V.P.$_2$ through A till it cuts the G.L. ? No! because A is not on a horizontal plane. V.P.$_2$ is on the horizon, and can only affect horizontal lines, and points in horizontal planes.

We have, therefore, to find a point on a horizontal plane which shall be in the same vertical plane as A.

Now, A.V.P.$_1$ is over V.P.$_2$; it is in the same vertical plane. And point A is in the same plane as A.V.P.$_1$, for A.V.P.$_1$ is in both the vertical and oblique planes, and A is in the oblique only. A.V.P.$_1$ is thus the point of connection between the oblique and vertical planes, as V.P.$_2$ is the point of connection between the vertical and horizontal.

We proceed thus. We bring a line down from A.V.P.$_1$ through A to the top front edge of the barge. From the point on the upper edge thus found, we take a line to A.V.P.$_2$, which transfers the point to the lower edge of the barge, which lower edge happens to be on the ground. Since A.V.P.$_2$ is immediately under A.V.P.$_1$,

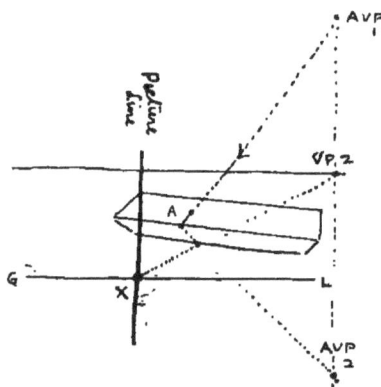

FIG. 154.—The picture-line found.

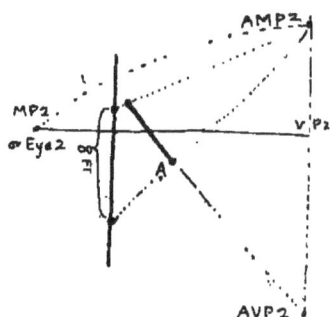

FIG. 155.—The mast drawn and measured.

the lines thus drawn zigzagging down the edge of the barge are under one another. We have thus a point found on the lower edge of the barge; it is also in the ground-plane; a line, therefore, from V.P.$_2$ through this point down to G.L., will carry the vertical plane through to G.L. The zigzag line drawn is really the trace of the vertical plane containing the mast upon the obliquely ascending top of the barge, the obliquely descending side of it, and the horizontal plane.

Through the point thus found on the G.L. we draw our vertical picture-line.

To measure the line is easy. We have the V.P.— A.V.P.$_2$; we have the eye (at M.P.$_2$). We thus describe an arc cutting the vertical V.L. at A.M.P.$_2$, and with it measure our 8 feet upward from A, along the new picture-line. See Fig. 155.

### QUESTION 5 (April, 1901).

*Diagram Q. 5 shows in perspective a vertical wall, and at its foot a point A on the ground-plane. C.V., E is the distance of the picture from the eye, which is 5 feet above the ground-line G.L. From a point on the wall 6 feet vertically over A project upwards a line 5 feet long to represent a post lying in a vertical plane perpendicular to the plane of the wall, the line making an angle of 30° with the ground-plane. The upper end of the post is to bear a sign-board, 2 feet square, with the device shown on the small diagram.* (30 marks.)

*The horizon must be drawn across the short way of the paper, 11 inches from the top, and the centre of vision placed 5 inches from the right-hand edge of the paper.*

The student who has been able to work Question 4 will have little difficulty with this. It is an example of an object in a *vertical* plane, and is simply an elaboration of the working required for finding the mast in the last question.

The whole of the scaffolding of the problem is given. The only omission is the tacit statement of the scale. We are told, however, that the eye is 5 feet above the

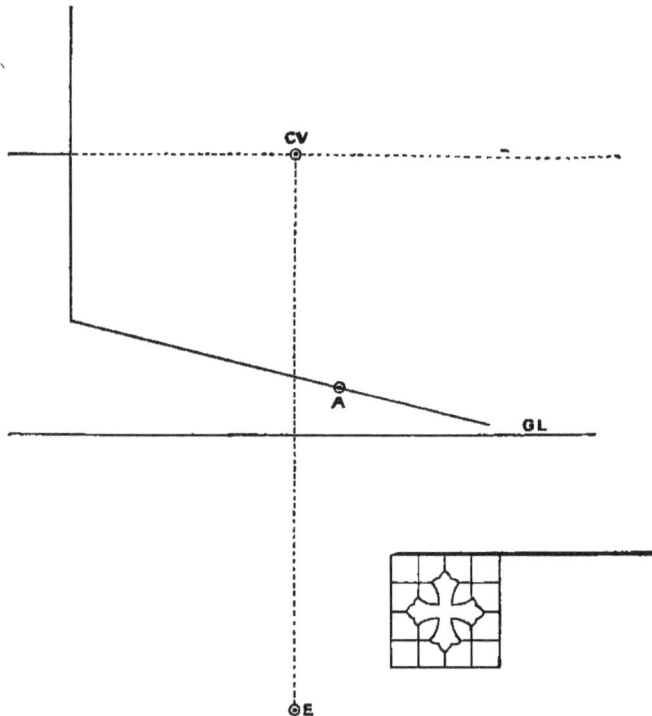

Fɪɢ. 156.—Diagram Q. 5.   (Scale ¼-inch to 1 foot.)

ground-line, and we consequently know that from H.L. to G.L. is 5 feet by scale.   If we divide the distance from H.L. to G.L. into five parts, we obtain our scale.   On the diagram this distance is $2\frac{1}{2}$ inches, so the scale is $\frac{1}{2}$ inch to 1 foot.   The reproduction here given is half that scale.

Before the real business of the question starts, we

have to find a point 6 feet above A. This new point is the real starting-point. To obtain this point we use the wall as a "wall of heights." We therefore continue the bottom line of the wall to the left till it cuts H.L. This is its V.P., and we continue it down to G.L., where we erect a line of heights which is really a vertical picture-line. Up this line of heights we mark 6 feet, and take a line back from that height to the V.P. From

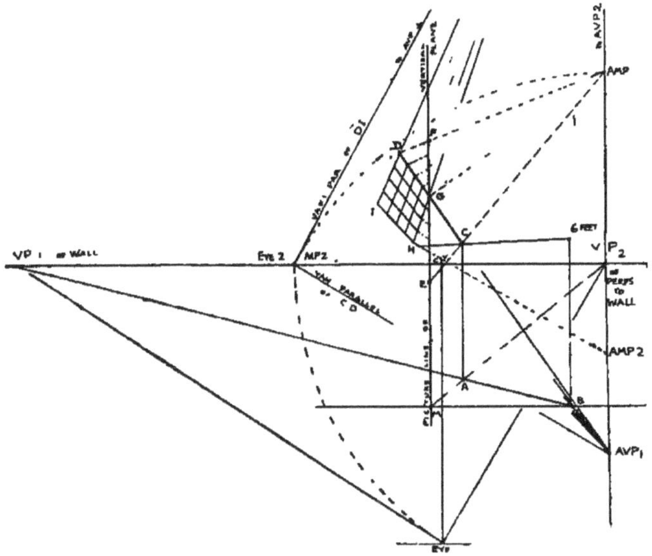

FIG. 157.—The sign in perspective.

A we now raise a vertical line till it cut this 6-foot line, and so obtain our point C.

We now proceed with the post or pole which carries the sign. It projects from the wall. It is in a vertical plane perpendicular to the wall. Before we go any further, therefore, we must find the V.L. and P.L. of such a vertical plane.

We have found that our wall vanishes to V.P.$_1$ on the left. We draw a vanishing parallel for V.P.$_1$ from V.P.$_1$ to the eye. This gives us the actual angle to the picture at which the wall is receding to the left. From the eye we therefore get another parallel at right angles to this first parallel, and this new parallel will give us V.P.$_2$ on the right. V.P.$_2$ is the V.P. of (horizontal) lines at right angles to (horizontal) lines vanishing to V.P.$_1$.

A V.L. vertically drawn through V.P.$_2$ will be the V.L. of a vertical plane at right angles to the wall. This V.L. will be the V.L. of the plane in which the post and sign lie.

To find the picture-line we look for a point on the V.L., which is also in the H.L., for our only picture-line so far is G.L., and it will be our guide to any new picture-lines. V.P.$_2$ is the point on both V.L.'s, so we draw a line from it through A, which is on the ground-plane, till it cut G.L., where we find the position of one point in our new vertical picture-line. We draw this P.L., therefore, vertically through the point just found.

The eye in relation to the vertical V.L. will now be required; it will be the M.P. of V.P.$_2$.

From the new eye we set off whatever angles we require. The post is at 30° to the ground, that is, to the horizontal. The flag-like sign will have some of its lines at right angles to this; that is, at 60° to the horizontal.

Does the post project upwards or downwards?

Upwards, because the statement further on speaks of the *upper end* bearing the sign. If the post project upward, and the upper end be the nearer, as is evidently the case, then the post, so far as the spectator is concerned, is vanishing *downward*. Its further end is lower than its nearer.

To find the V.P., therefore, we set off 30° downward from the horizon; for in this case the line from eye$_2$ to V.P.$_2$ represents the level, and happens to be part of the horizon. 60° upwards will find the V.P. of the shorter lines of the sign. For both of these A.V.P.'s we find the A.M.P.'s.

It is not necessary to follow out the measuring and drawing of the different lines of post and sign.

### QUESTION 6 (April, 1901).

*Diagram Q. 6 represents in perspective a wall composed of vertical and oblique planes, and two crosses, the one on the right standing on the ground-plane, the other raised on a rectangular block which also stands on the ground-plane. The position of the horizon, the centre of vision, and the distance of the eye from the picture are given. Draw the shadows cast by these objects, the sun being behind the spectator in a vertical plane receding to the right and inclined at 60° with the picture-plane, the altitude being 30°.*          (32 marks.)

*The horizon must be drawn across the short way of the paper, 7½ inches from the top, and the centre of vision placed 5½ inches from the right-hand edge of the paper.*

The sun is behind the spectator, and therefore will not be seen. Its rays will pour downward, and will consequently seem to vanish down somewhere below the horizon.

The sun is behind the spectator on the left, but its rays vanish in planes at 60° to picture-plane towards right. The V.P. of sun's rays will be down somewhere on the right.

We find the V.P. of sun's rays precisely as we found the V.P. of the post in the last question. It is a V.P. in

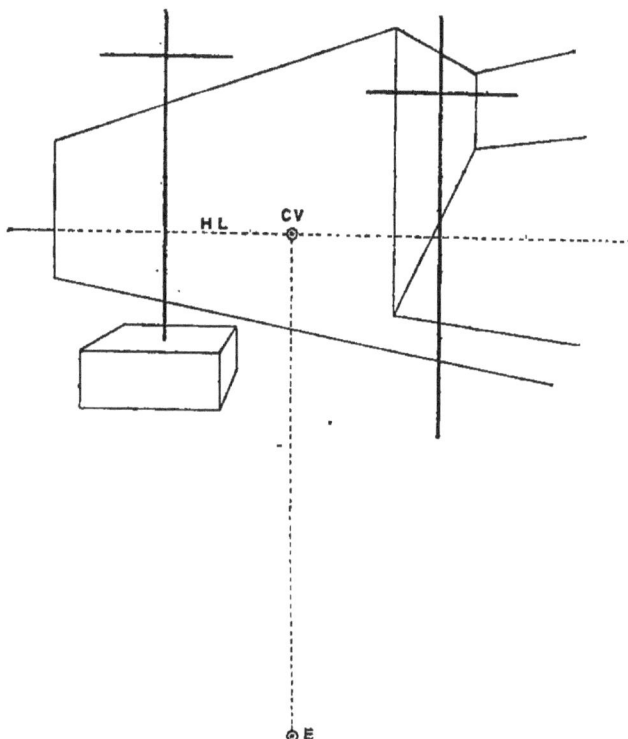

FIG. 158.—Diagram Q. 6.

a vertical V.L. From the eye we therefore set off a vanishing parallel in the usual way at 60° to the right. This gives us a V.P. on the horizon, V.S.₁. Down through this V.P. we draw a vertical V.L., and find the eye in relation to it—the M.P. of the V.P. From this new eye we set *down* 30° to the horizontal (not from a vertical directing line through the eye—the 30° is to the level, and not to the upright position).

To find shadows we always require the V.P. of the shadows—V.S. it is called. The V.S. of the shadow of a line changes with each plane the shadow falls on. The rule is simple : Join V.P. of line to V.P. of sun, and where this line cuts V.L. of plane receiving shade is the V.S. required. Let us proceed to find the V.S. or V.S.'s of our vertical lines.

Through V.P. of sun s rays we therefore draw a long vertical line—we have it already. Where does it cut H.L. ? For there is the V.S. (V.S.$_1$) for the ground-plane. From the foot of the cross on the right we draw a shadow to V.S.$_1$. We find our shadow crosses the bottom line of the wall; we suspect it was intended to do so. We therefore have to run the shadow up the wall from the point where it strikes the base of it.

Before we attempt to get the shadow on the wall, let us find the V.L. of the wall, for the wall is a plane receiving shade, and V.S. of shadow on it occurs on its V.L. We find sufficient is given of the wall to enable us to find its V.P. on the left. Through the V.P. we draw a vertical V.L., which is the V.L. of the plane of the wall. When we proceed to get the shadow of our cross on the wall, we see that our cross consists of two lines, in two directions. One is vertical, and so has no V.P. ; the other, the arms, is quite level,—it too must be parallel to the picture, and has no V.P.

When lines have no V.P., we cannot join their V.P.

to the sun to get their V.S., but we *draw from the sun parallel to them.*

Hence, while the line from the sun parallel to the shaft

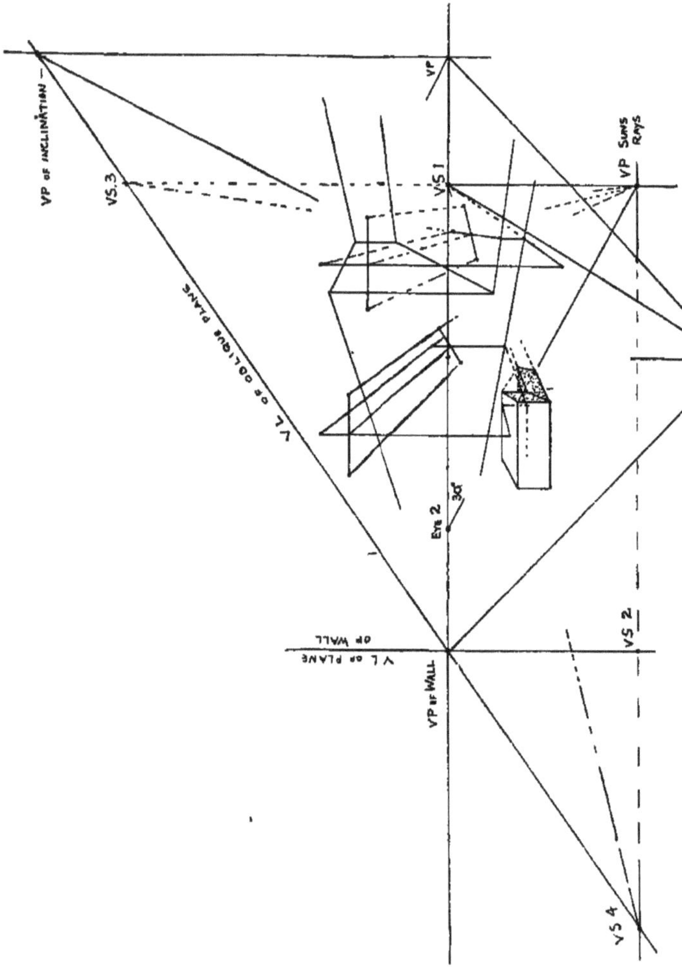

FIG. 159.—The shadows projected.

of the cross will never meet the V.L. of the wall, the line parallel to the arms will at V.S.₂.

The shadow of the shaft will have to remain parallel to

the line which cast it, as no V.S. can be found; but the arms will vanish to V.S.₂. But a further plane confronts us. Our wall has not received the shadow of the cross when an oblique plane interferes. We know this inclined plane is oblique, because its horizontal trace does not go to C.V., but vanishes to the left.

We have to find the V.L. of the oblique plane. The V.P. of intersection will be the V.P. of the wall on the left. We shall have, by means of the eye, to get the V.P. at right angles to this, on the right. Through this new V.P. we raise a vertical, and continue the slanting line of the oblique plane to it. Obviously this slanting line is in a plane perpendicular to the wall, and will thus give us the V.P. of inclination. Through the V.P. of intersection on left, and this V.P. of inclination high up on the right, we draw our V.L. of the oblique plane, drawing it well past the V.P.'s used to obtain it. We see that our two parallels from the sun, one vertical, the other horizontal, strike this slanting V.L. as they did the others. We thus obtain V.S.₃ and V.S.₄, and use them in the same way as the others.

The block on which the second cross stands will give little trouble. Its V.S.'s are V.S.₁ and the C.V.

## Section (B).

The following General Instructions precede the questions:—

*Read the General Instructions on page 182.*
*You may attempt three questions only. In the solution*

*of these accuracy of measurement and use of instruments
are not demanded, so long as the method of obtaining the
required result be clearly and concisely indicated, and the
working points and lines marked with their respective
explanatory letters and, where necessary, angles with
figures.*

*The placing on the paper of the horizon and centre of
vision is not prescribed. The candidate can readily ascertain from the diagrams the space the working lines will
occupy on his paper.*

### QUESTION 7 (April, 1901).

*Diagram Q. 7 shows in perspective a rectangular solid*

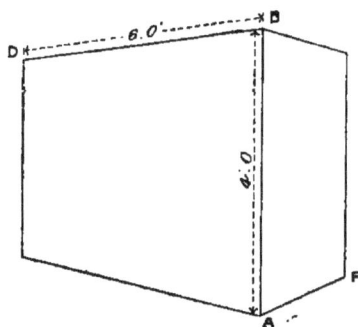

FIG. 160.—Diagram Q. 7.

*standing on a horizontal plane.
Edge AB is known to be 4
feet high, and BD 6 feet
long. Find the vanishing
points, horizon, measuring
points, distance of the eye,
and centre of vision, also the
length of AF. (6 marks.)*

The problem is an instance
of working backwards to the scaffolding.

The vertical line AB is known to be 4 feet. If we
draw our picture-line through A, we make AB actual
size, and it becomes our scale. AB is the junction of
two surfaces receding to right and left. If we continue
the receding lines, we shall obtain the two V.P.'s. We
draw our horizon through them, and must be careful that

it is at right angles to AB, which is vertical, and also
that our picture-line is parallel to the horizon. To find
the eye and C.V. we proceed as follows : We know that
two reciprocal V.P.'s have their vanishing parallels
enclosing a right angle. We know that the right angle
occurs at the eye, and we know that in geometry a
semicircle is such that any two lines from the extremities
of the diameter to a point in the curve contain or
enclose a right angle. Taking the distance on our

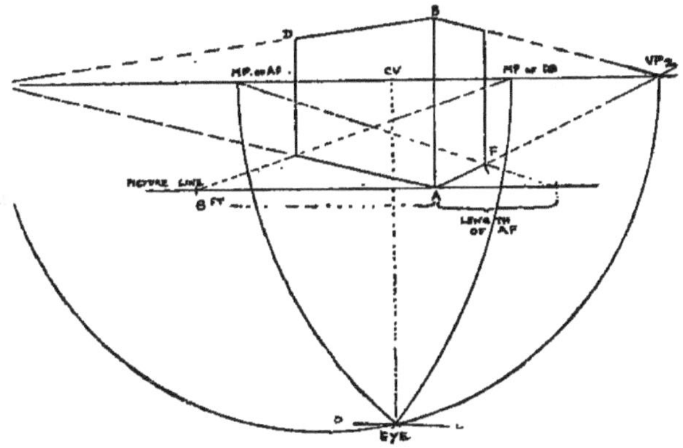

Fig. 161.—The problem solved.

horizon from one V.P. to the other as the diameter
of a semicircle, we first bisect the distance, and then
describe the semicircle. Somewhere upon this semicircle the position of the eye will occur. If we did
not know the length of one of the receding sides, we
could not find the position of the eye. If the object
were a cube, we should know that the receding sides
were the same in size as the upright edge, and, therefore,
in that case no dimension other than the front edge need
be given. In the case before us, however, the object,

though cuboid, is not a cube. We are told DB is 6 feet. Knowing this, we find the M.P. of the V.P. on the left by putting 6 feet along our picture-line from A toward the left, as though we were going to measure 6 feet along the edge of object. Then, instead of drawing from the limit of 6 feet to the M.P., we draw from it through D (or rather the point below D on the lower line), and this line taken up to H.L. will give the M.P. of the V.P. of DB.

Now, the M.P. is the same distance from its V.P. as the eye is from the V.P., so we describe an arc with V.P. as centre, and distance V.P. to M.P. as radius, and cut the semicircle. This gives the position of the eye. The C.V. will be vertically above it.

Having found the eye, we can get the M.P. of the V.P. of AF, and so can measure AF on the picture-line.

### QUESTION 8 (April, 1901).

*Diagram Q. 8 shows in perspective a circular dish standing on a horizontal plane. The position of the horizon, the centre of vision, and the distance of the eye from the picture are given. Show how the aspect of the dish would be changed, if it were raised vertically the distance of AB the bottom plane of the dish remaining horizontal.*

(8 marks.)

Sketch around the lower circle of which A is the centre a square such as is used for obtaining circles in perspective (see page 220). The sides of the square may be two of them parallel to H.L., and the other two vanish at C.V. Make a similar square around point B, the corners of the square of B being exactly over the corners of the square of A. Within it draw the circle. Make in the same way a square around the larger circle below, and construct another at the proper distance above B. This second large circle must, of course, have its corners exactly over the former large square. To find how far above B the upper circle should be, draw across the larger circle below a horizontal diameter; this will give the centre of the large circle where it crosses the dotted line.

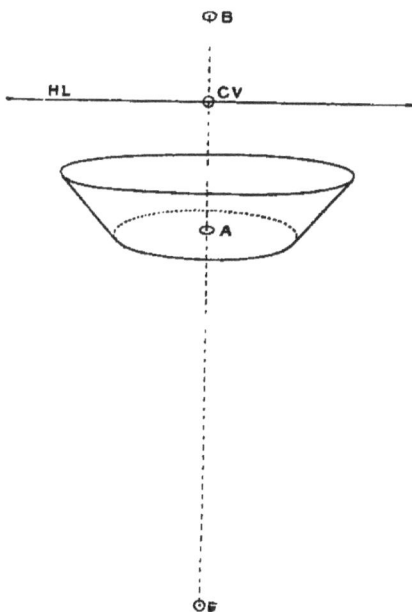

Fig. 162.--Diagram Q. 8.

Now, vertical measurements are not subject to diminution; therefore the centre of the upper large circle will be precisely as far above B as the lower circle is above A.

The distance of the eye being given, the problem could be worked by parallel perspective; but the readiest way of working, even in a sketchy way, is to get the D.P.'s by setting off the distance C.V. to eye along H.L. from C.V., and using D.P.'s as V.P.'s of diagonals. Thus we know that diagonal GJ will vanish to same V.P. as diagonal CF. We proceed thus: Having drawn lines of indefinite length to represent CD and EF, we bring diagonals through A to cut the points CD and

EF. CE and DF should then vanish to C.V. Then, with the circle above, we commence by a diagonal through B. We raise verticals from C to cut ' the diagonal in G, and from F to cut the diagonal in J. With G and J found, we can readily complete the square around B. The upper or large circle has now to be repeated above. We are not given its centre, so have to find it; a line across the oval is not reliable, though roughly it will serve well enough. We have the front

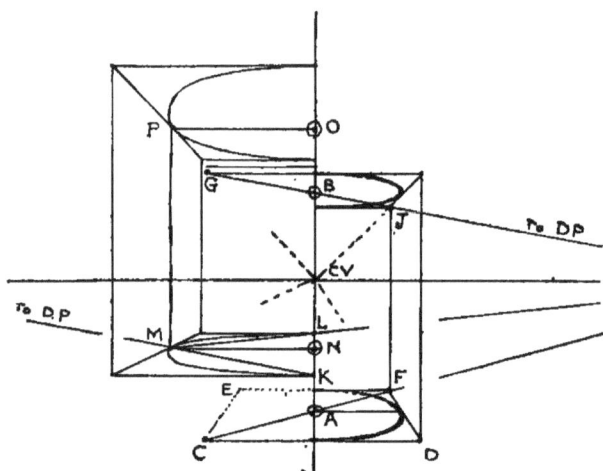

Fig. 163.—The solution of the circles

and back points K and L. By drawing lines to the V.P.'s of diagonals (D.P.'s), we obtain KM and LM, with M as the centre of the left side. From M we can readily find N the centre, and can complete the square. We now make BO equal to AN, because the axes are equal and vertical; then P out sideways, cut by a vertical from M. With P found, we can easily complete the square.

QUESTION 9 (April, 1901).

*Diagram Q. 9 is a perspective view of a rectangular pier supporting the remains of a semicircular arch, the springing-line of the arch being at level Sp. On the right is shown the trace on the ground of the other pier. The horizon, the centre of vision, and the distance of the picture from the eye are given. Restore the whole archway.* (12 marks.)

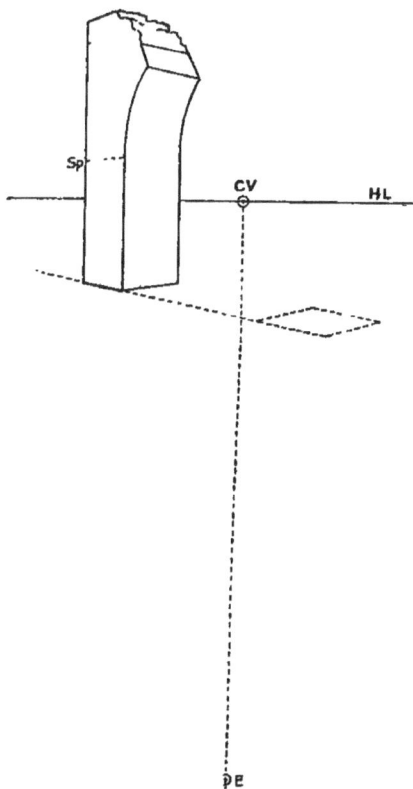

Fig. 164.—Diagram Q. 9.

Here we continue the trace of the face of the arch on ground to the left, and there find a V.P.$_1$.

Having the eye given, we can find V.P.$_2$ on the right, the vanishing parallels being at right angles. We can also easily find the M.P.'s, and we can draw the G.L. through the nearest (lowest) point of the dotted plan of the pier on the right.

From the dotted plan we raise four vertical lines to form the corners, and limit them above by lines to V.P.$_1$ and V.P.$_2$ in continuation of the portion left of the top.

If we care to guess the curve of the arch, we can complete the drawing without further working.

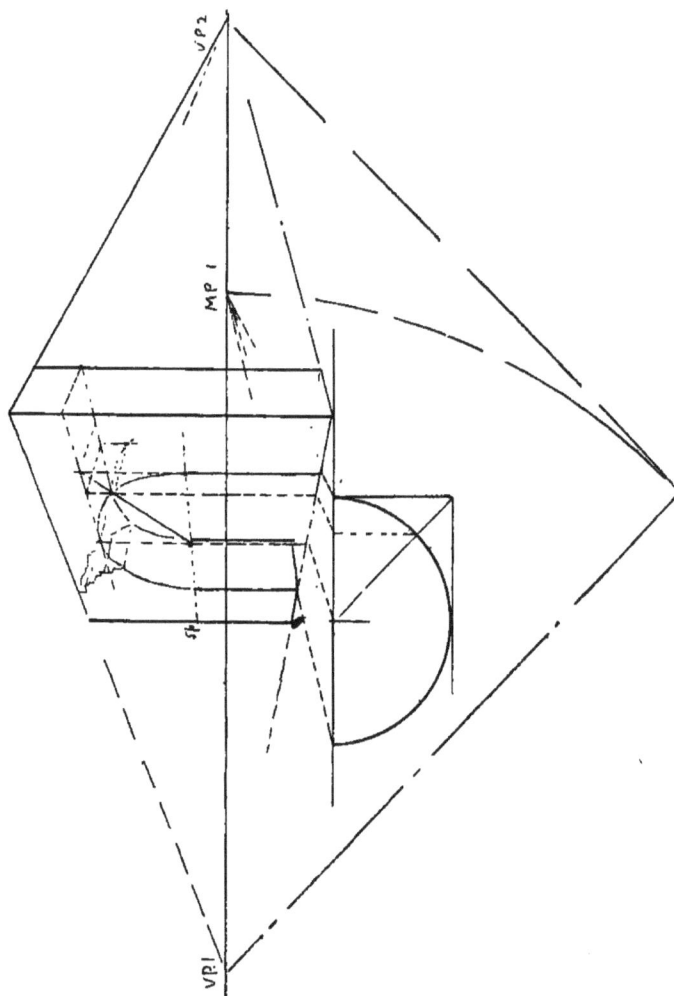

Fig. 165.—The arch restored.

If not, we have to find the exact size of the semi-circle. By M.P.₁ we can find the size on the G.L. of the span. The span is the diameter of the semicircle.

We therefore readily complete it by the method seen in Fig. 165.

## QUESTION 10 (April, 1901).

*Diagram Q. 10 gives a perspective view of a road descending from the ground-plane, HL being the horizon. At point A, which is on the ground-plane, erect a vertical*

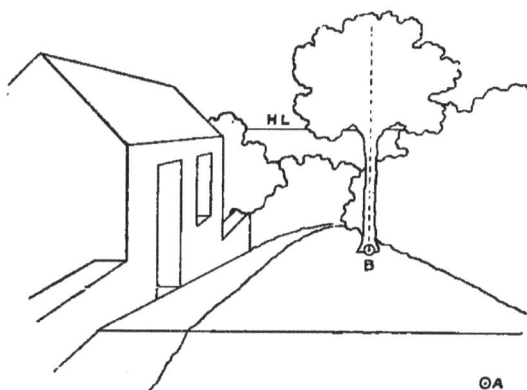

FIG. 166 —Diagram Q. 10.

*line to represent a man, equal in height to the nearest edge of the cottage door, which is assumed to be 6 feet. Find also the height of the tree standing at point B.* (8 marks.)

Apparently the house vanishes to C.V. This is evident from the short side of door-step, window-sill, and edge of first wall being level, and not vanishing. The line across the road above A also indicates this 'fact.

We may then first transfer A across to the line where the wall touches the G.P. on the left, C. This we do with a level line, since, being in apparently parallel perspective, the wall is perpendicular to the picture.

Having found this level of A on the foot of the wall, and having found the V.P. of the house, we draw a line from V.P. through the top of the door, and meet this line by a vertical from C at foot of the wall. CD is not the height of the door, because the door is actually below the level of C. We continue the vertical DC downward a little, and then draw another line from V.P. through the lower end of the front line of the door; and where

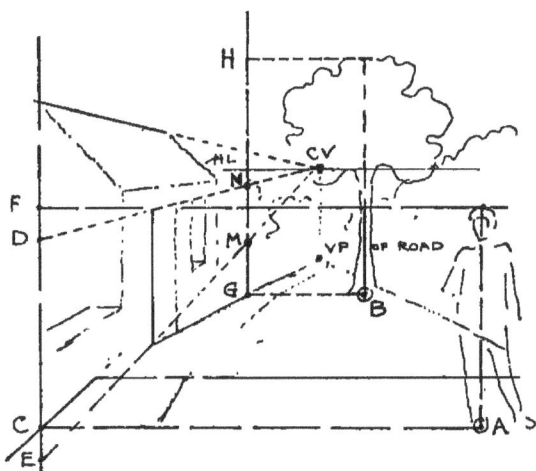

FIG. 167.—The problem solved.

this cuts CD in E is the level of the door below the G.P. ED is the height of the door.

Now, the man at A is to be the same height, but on a different level. Take distance CE, and put it above D to F. FDC and E are all on the same vertical line, and FC is equal to DE. Transfer by a level line height F across to A.

To get the height of the tree. Presumably the road is level from side to side though descending. We can transfer B by a horizontal line to the lower edge of the cottage. The point is just where the straight line ends. At G we erect a vertical, and transfer the top of the tree to it, H. GH is the height of the tree, but some distance back. GH is crossed by lines from D and E to CV, representing between them 6 feet. The distance MN is therefore 6 feet. If so, hⱼw high is GH? About 24 feet. One has simply to prick off size MN up GH or divide MN into six parts, and thus make a scale.

### QUESTION 11 (April, 1901).

*Diagram Q. 11 gives in perspective a group of models (copied from an exercise submitted by a candidate in an examination in model drawing), consisting of a board which supports a hexagonal prism, a square pyramid, and a white bottle. The correct position of these objects is shown by the small plan. Find some of the errors in the perspective and indicate the methods for rectifying them. The position of the horizon, and of the centre of vision, and the distance of the eye from the picture are to be as assumed in the diagram.* (12 marks.)

Question 11 could be solved entirely with correct measurements if we assume, say, the first corner of the board to be in the proper place below the eye and touching the picture. Such a thorough setting straight is not asked for. V.P.$_1$ for the front edge of the board, V.P.$_2$ for the sides, V.P.$_3$ for the face of the prism, and

V.P.$_4$ for the side of it are necessary, and can be obtained by vanishing parallels from the eye, taking the angles from the plan. With these four V.P.'s a great, deal of correction can be done. The mistakes in the diagram

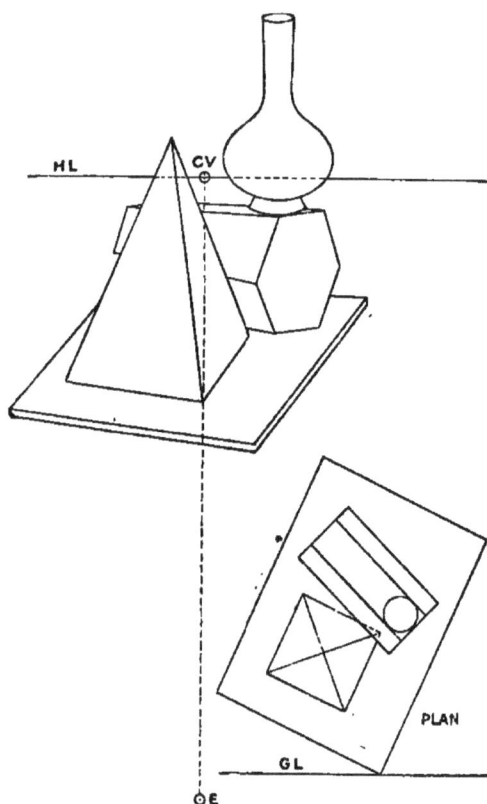

FIG. 168 —Diagram Q. 11.

are fairly numerous. Of course the mouth of the vase should not be visible, and the foot is much too curved, and does not form a circle lying on the top of the prism as it should.

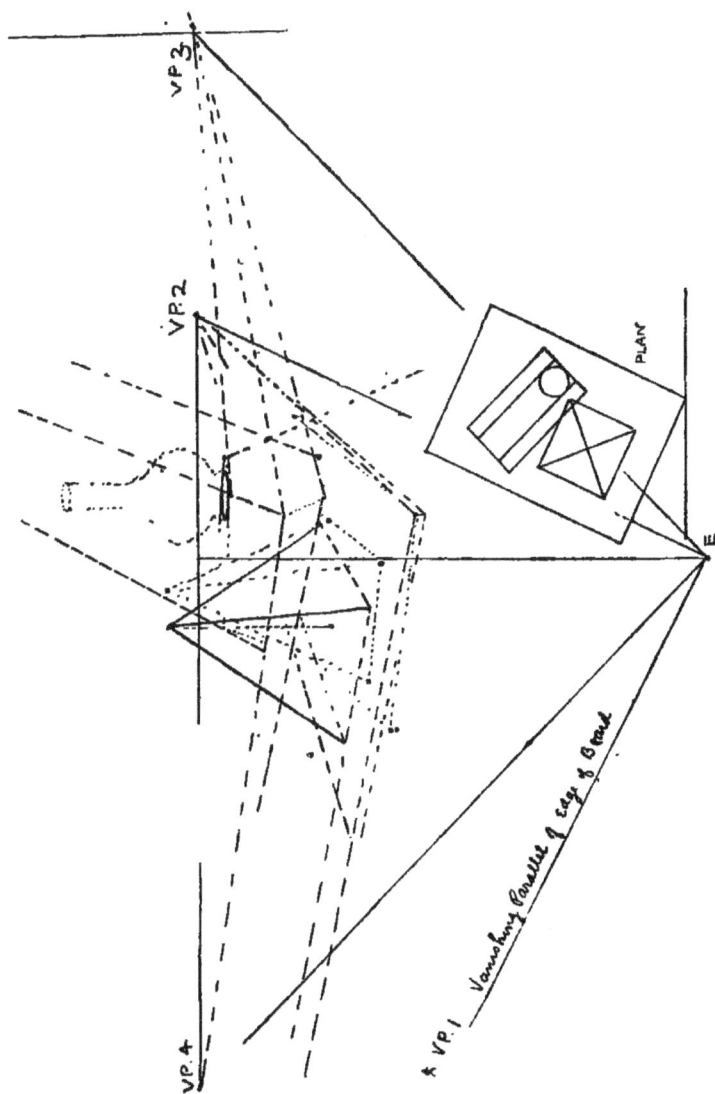

FIG. 169.—The influence of the V.P.'s

VP 3

VP2

VP 4

* VP.1 *Vanishing Parallel & edge of Board*

PLAN

E

QUESTION 12 (April, 1901).

*Diagram Q. 12 gives in perspective a vertical surface standing on a horizontal plane. AB is a vertical line erected on the same plane and casts the shadow shown by the dotted line, as does the line DF which is perpendicular to the vertical plane. The horizon, the centre of vision, and the distance of the eye from the picture are given. Find the source of the artificial light and draw the shadow of the line HI which is parallel to DF.*

(14 marks.)

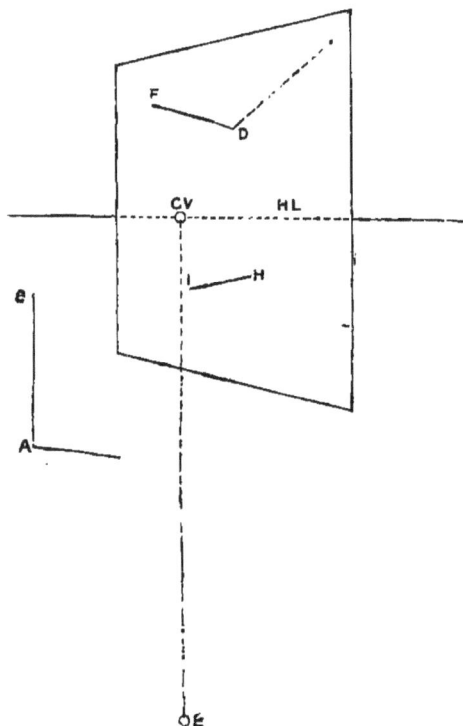

FIG. 170.—Diagram Q. 12.

The student will remember that shadows cast by artificial light *radiate* from points on the surface on which the shadow exists, that this point of radiation is found by drawing a line from the light parallel, in fact, to the line whose shadow is required till it strikes the plane receiving the shadow. He will also remember that the light (whether sun or artificial) is used directly only as a means of measuring the shadows. The rays which measure shadows are drawn from the light. Now, the rays which measure the shadows given in this diagram will pass through the ends of the lines B and F, and the ends of the shadows. If these rays be both drawn they will give, where they meet the position of the light, L.

If we were now finding the shadow of DF, we should first find the radiating point of the shadow on the vertical plane. The line DF is perpendicular to the vertical plane; its V.P. will, therefore, be somewhere on the horizon. Continue FD down to HL and we obtain its V.P. This

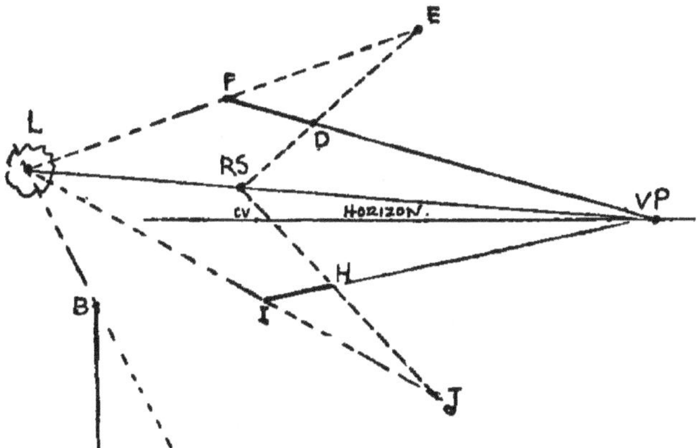

FIG. 171.—The problem solved.

V.P. will also be the V.P. of all lines parallel to DF. To obtain our radiating point of the shadow of DF we have to draw from the light a line parallel to DF. This line will go to V.P., which is the V.P. of all lines parallel to DF. This line from L to V.P. will cut the vertical plane somewhere above C.V. We should always need to be told in what manner the light were related either to the vertical plane or the ground. This indication is provided in this case by the shadow of DF being given. If this shadow is drawn from a point somewhere on the line from L to V.P., we have only to continue the line of the shadow and we shall obtain the point, R.S.

This R.S. will also be the radiation point of the shadow of HI, because HI is parallel to DF. We therefore draw from R.S. through H a shadow, and cut it off by a ray from the light through I.

### QUESTION 1 (June, 1901).

*Diagram Q. 1 gives the plan and elevation of a rect-angular slab lying on the ground-plane with its sides inclined to the picture-plane at angles of 60° and 30°. Put the slab, and the device shown on its upper surface, into perspective. The point A is to be on the ground-line 1 foot to the left of the spectator. The eye is to be 6 feet above the ground and 11 feet from the picture. Scale, ½ an inch to a foot.* (13 marks.)

*The horizon must be drawn across the long way of the paper, 2½ inches from the top, and the centre of vision placed on the centre of the horizon.*

The subject is a square slab with a pattern upon the upper surface. The only danger or "catch" can be in the pattern being on the upper surface, for the student may make the slip of measuring the positions of the points of the star from the ground-line direct to the upper edge

of the slab. This is the fault already spoken of in
Question 1, p. 189.

*Elevation*

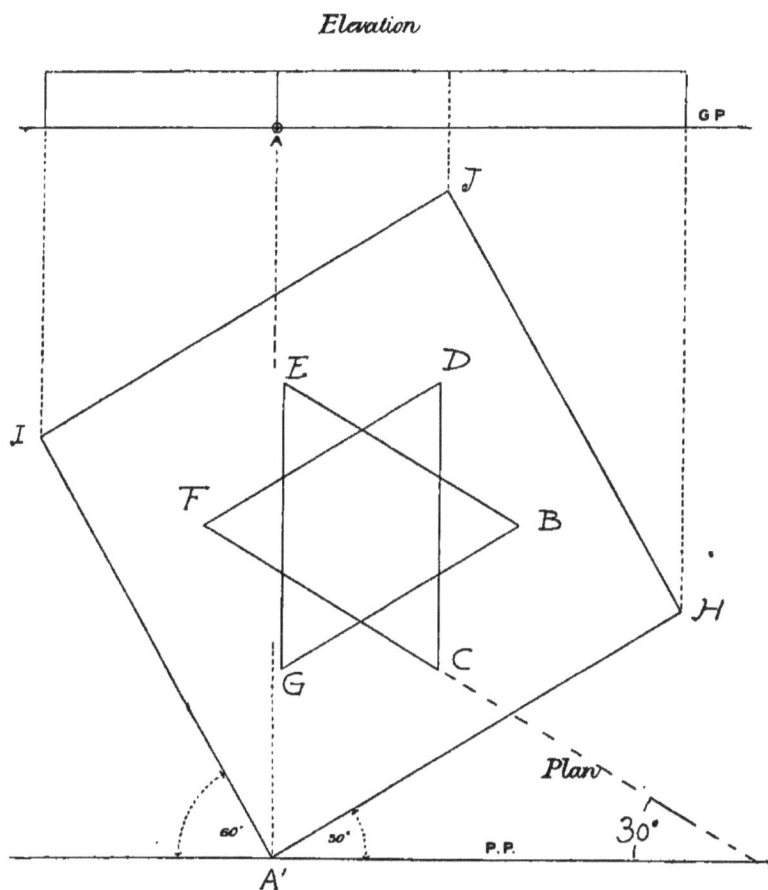

FIG. 172.—Diagram Q. 1. (Scale ¼ inch to 1 foot.)

Excepting for this, the problem is one of very straight
sailing.

The ground-line will be 6 feet by scale down below
the horizon. The picture-line for the top of the slab

would be a shade over 1 foot above G.L. Therefore, in
Fig. 173, A up to A' is just over 1 foot.

The plan shows AH vanishing at 30° toward the right,

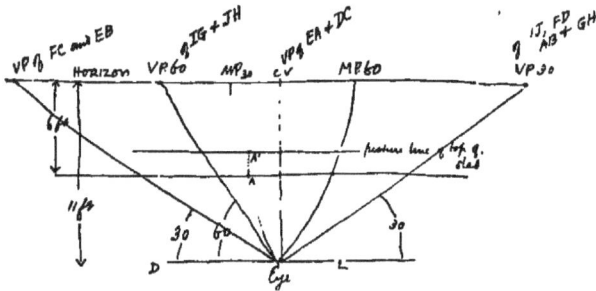

FIG. 173.—The scaffolding of the problem.

IJ also, therefore, vanishes at 30° toward the right.
while IA and JH vanish at 60° toward the left. Of the

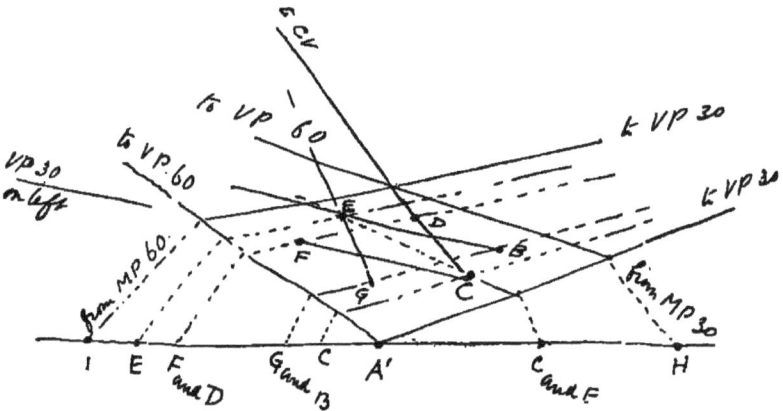

FIG. 174.—Sketch of the working. Note that the picture-line is not
the ground-line.

lines of the star, two vanish to the V.P. 30° on the right.
These are GB and FD. Two, FC and EB, make 30° with
the picture toward the left, while DC and EG, both at

60° to GB, are perpendicular to the picture, and will vanish at C.V.

The eye is placed 11 feet down from the horizon (*not* from the ground-line), the distance from C.V. to eye being 11 feet by scale.

The whole of the V.P.'s being easily obtained, it will be well to use them for finding the star, and not to merely find the positions of B, C, D, etc., by parallels to AH and AI.

### QUESTION 2 (June, 1901).

*Diagram Q. 2 shows the half plan of a square ceiling. Put the whole ceiling into perspective with its nearest corner A 12 feet vertically over a point on the ground-plane 3*

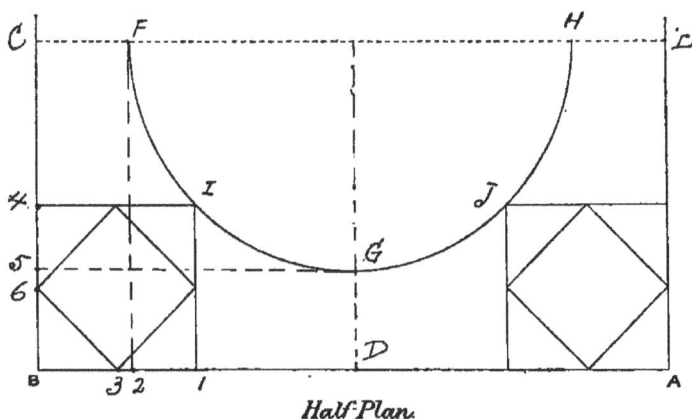

*Half Plan.*

FIG. 175.—Diagram Q. 2. (Scale ¼ inch to 1 foot.)

*feet to the right of the spectator and 1 foot within the picture. The edge AB is to be inclined to the picture-plane at an angle of 30° towards the left. The eye is 12 feet*

232 *Perspective*

*from the picture and 5 feet above the ground. Scale, ½ an inch to a foot.* (15 marks.)

*The horizon must be drawn across the long way of the paper, 6 inches from the top, and the centre of vision placed 9 inches from the right-hand edge of the paper.*

The problem requires (1) the understanding of a plane above the ground, (2) the putting into perspective of a square, and (3) the striping of the square to enable a certain pattern to be drawn.

The point A, which is 3 feet on the right and 1 foot within the picture, can be either found on the ground and then raised by a line of height 12 feet, or it may be

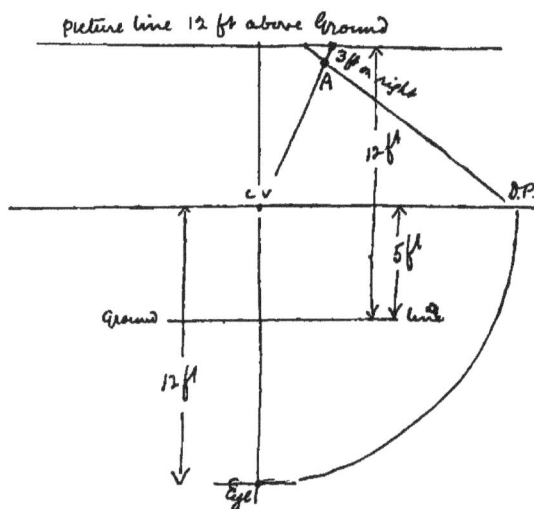

Fig. 176.—The scaffolding of the problem.

found at once on a plane 12 feet above the ground. Indeed, the ground-line is not wanted at all. The relations of the picture-line, horizon and ground-line, as well as the method of finding point A, are shown on Fig. 176.

The plan has been lettered for reference, and the dotted lines which give points D, G, 1, 2, 4, and 5 have been drawn upon it. The square has thus several lettered or figured points on each side. Thus AB has D in the centre, and between D and A points 1, 2, and 3, and between D and B also points 1, 2, and 3. The dimensions on the other sides are the same as those on side AB.

To work the problem, we draw from A a line to V.P. 30 on left, and one to V.P. 60 on right. Along both these we measure the dimensions just spoken of. Taking the line to V.P. 30, we commence by bringing a line from M.P. 30 through A to A' on the picture-line. Along the

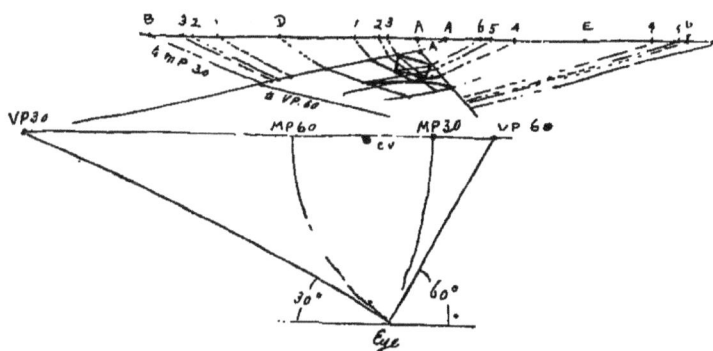

Fig. 177.—Sketch of the working.

picture-line toward the left, we set off points 3, 2, 1, D, 1, 2, 3, B, transferring the dimensions from the plan. We then draw lines from all these points to M.P. 30, and note where they cut the line from A to V.P. 30. We thus have the side AB with all its measurements upon it.

We do the same with the side vanishing to V.P. 60. When we have the two sides marked out in this way, we proceed to stripe the square by lines from all these points to either V.P. 30 or V.P. 60. The striping giving us all the points of the pattern, we draw in the circle by freehand, and rule in the small squares in the corners.

The exercise seems to be one rather for careful working than for particular knowledge. If there is one fault more than another against which the student should be warned, it is using lines going to either M.P. after they have crossed the side of the square they are supposed to measure. It is well not to take the measuring lines past the lines they are intended to measure.

### QUESTION 3 (June, 1901).

*Find a point on the ground-plane 2 feet to the right of the spectator and 12 feet within the picture. This point is to be the centre of a circle of 16 feet diameter, which is the base of a circular platform 1 foot high. Draw the platform, and on its centre place a cube of 8 feet edge, having its right-hand side inclined to the picture-plane at an angle of 30°. The upper surface of the cube is to be the base of a pyramid 5 feet high. The eye is 10 feet from the picture and 5 feet above the ground. Scale, ½ inch to a foot.* (20 marks.)

*The horizon must be drawn across the short way of the paper, 6 inches from the top, and the centre of vision placed 5¾ inches from the left-hand edge of the paper.*

The ground-line is 5 feet below the horizon; but the plane upon which most work occurs is 1 foot above, that is, the plane of the top of the circular platform.

Not only have we to get the top of the platform, but

we have upon the top to place a cube. The cube having
nothing to do with the *bottom* of the platform, it will be
best to commence with the top of the platform, using a
picture-line 4 feet below the eye, and then dropping
down 1 foot to the bottom.

To get the top of the platform, we proceed as follows
(Fig. 178): We find 2 feet on the right along the
picture-line from the middle, and draw a line from 2 to

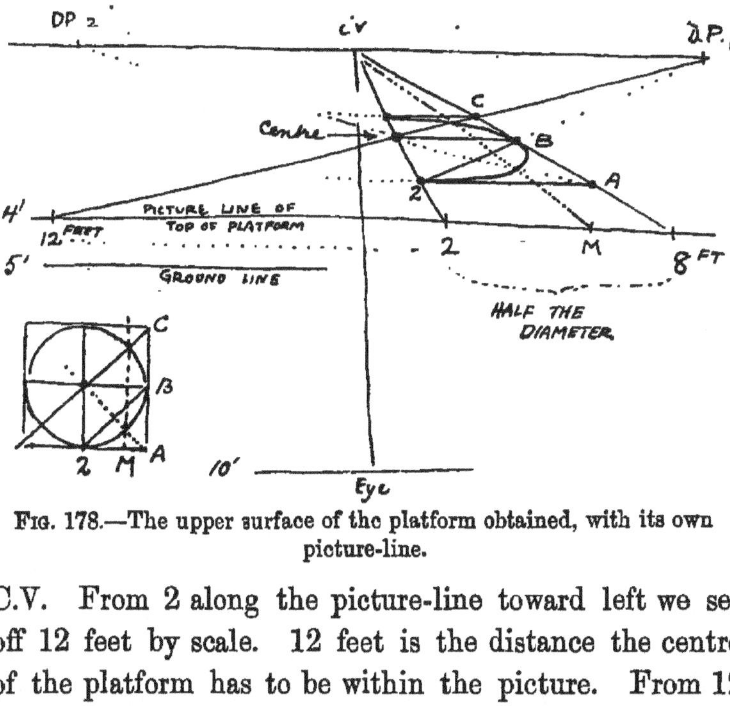

Fɪɢ. 178.—The upper surface of the platform obtained, with its own
picture-line.

C.V. From 2 along the picture-line toward left we set
off 12 feet by scale. 12 feet is the distance the centre
of the platform has to be within the picture. From 12
on the left we draw to D.P.₁ on the right, and this line,
by crossing the line from 2 to C.V., gives our centre.
It will also give us the diagonal ending in C, as seen in
the sketch-plan at the side.

The circle can be treated in parallel perspective, and

the D.P. thus gives us, as we have just seen, the V.P. of diagonals. The sketch-plan shows the scaffolding by which we get at our circle. Of this scaffold we have so far only the line through the centre to 2 and the diagonal to C, though we have not point C. We now get the side AC by setting off 8 feet (half the diameter) along P.L. from 2 toward the right. From 8 we draw to C.V. a line parallel to that from 2. Where this cuts the diagonal is point C. We now draw with the T-square the level line from C towards the left, and it gives the

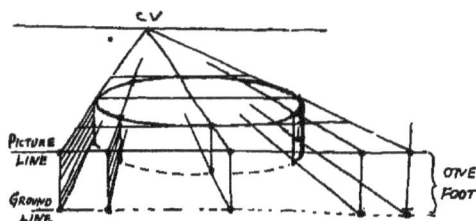

Fig. 179.—The lower edge of the platform obtained by lines of height dropped down 1 foot from the picture-line to the ground-line.

farthest point. From the centre we draw a level line toward the right, and it gives point B where it crosses the line 8C. The small diagonal B2 on the plan is next drawn from D.P.$_1$, and it gives us point 2. A level line from 2 gives A.- A diagonal to D.P.$_2$ through centre will contain the point in the circle which is located by the line M. To find distance 8 to M, we have to make a careful plan of the proper scale. One-quarter of the sketch-plan here given is sufficient, but the centre, B, A, 2, the quadrant, and the line M parallel to AB, must be all accurately found. Distance M8 is then set off from 8, and line M to C.V. drawn, giving two points in the curve where it crosses the diagonals. The curve must be drawn with grace, and will most probably extend beyond the point B, so that the greatest width will depend on the good drawing of the curve rather than the accurate perspective working of the lower curve of the platform; only the front need be drawn, though it is no harm to draw the whole. Still, neatness and clearness must be considered, for an examiner will hardly prefer a chaos of working lines, if the chaos is due to adding unnecessary parts. It is not stated by the examiner

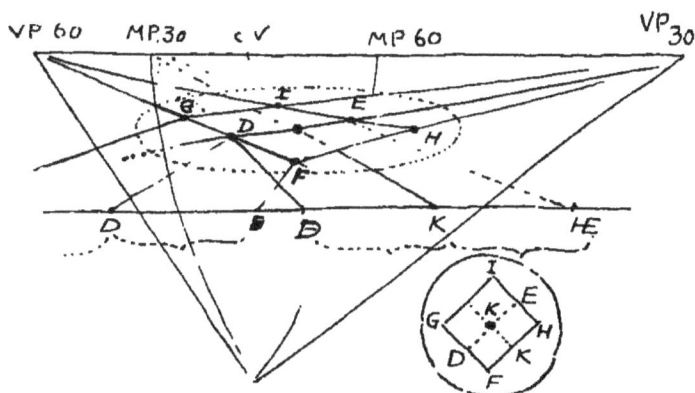

FIG. 180.—The square base of the cube found on the top of the circle.

whether he wishes the invisible parts put in by dots, but none of his diagrams have them.

The next item is to place a cube on the top of the platform, the axis of the cube presumably coinciding with the axis of the platform. We have the centre of the circle as our start-point K. Our sketch-plan is shown in the diagram. We shall have to add DE to our square, in order that we may get from K to the sides.

Our V.P.'s are on left and right, and their M.P.'s must

be found. The square is in the plane of the top of the platform, so the measurement will be made on the picture-line 4 feet below the eye. We proceed thus: We draw from V.P. 30 on right a line through the centre K. Then we measure along the line from K, KE and KD. To do this we bring a line from M.P. 30 down to the P.L., and so obtain point K on it. Then on either side we put the actual measurements KE and KD to right and left, and then take lines back to M.P. 30, and so obtain D and E.

We next go to V.P. 60, and draw lines down forward

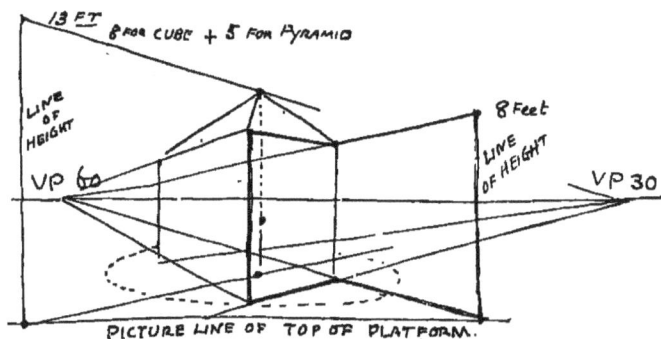

Fig. 181.—The cube and pyramid raised.

some little way through D and E; and along the one through D we measure DF and DG, using M.P. 60. F and G found, lines from them to V.P. 30 will complete the square.

To raise the cube to its proper height, and to raise above it the apex of the pyramid, is the work of but a few moments. It will be best not to use the same line of height for the apex as for the cube. The student must keep a sharp look out for heights such as of the pyramid, where the apex is not over the corner of the square. There is a temptation to which the student not unfrequently succumbs—to forget to work one's way in from the corner of the cube to the centre. That is to say, an extra 5 feet on the line of height on the right will *not* measure the height of the apex, because the wall of height for the corner does not go through the apex. The dotted axis of cube and pyramid must stretch from top to bottom of the wall of height for the apex.

Question 4 (June, 1901).

*Diagram Q. 4 shows in perspective a rectangle ABCD of no thickness, lying on the ground-plane, and a rect-*

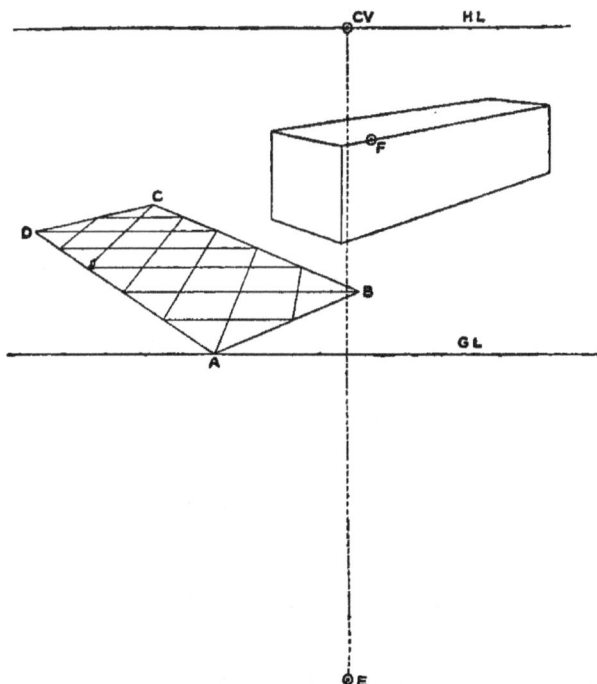

Fig. 182.—Diagram Q. 4

*angular block also lying on the ground-plane, on which is a point F. The horizon, centre of vision, distance of the eye from the picture, and the ground-line are given. Find the dimensions of the rectangle, and draw in perspective a similar rectangle of the same size and with the same pattern upon it. One of its shorter edges is to be on the ground-plane parallel to the longest edges of the block, and the longer edges of the rectangle are to make an angle of 30° with the ground, and to be in vertical planes parallel to the ends of the block. The left-hand edge of the rectangle is to pass through the point F.* (30 marks.)

*The horizon must be drawn across the short way of the paper, 7 inches from the top, and the centre of vision placed in the centre of the horizon.*

What is evidently wanted is such a result as is sketched in Fig. 183. How far H will be from G will be determined

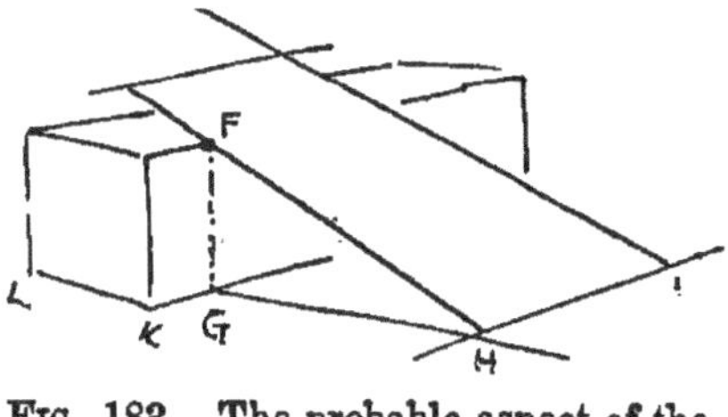

FIG. 183.—The probable aspect of the problem when solved.

by the fact that the angle FHG has to be 30°. Whether, therefore, the rectangle overhang the back, as shown in the sketch, will be revealed in the working.

The long edges FH, are to be in vertical planes, and the short edge HI is to be parallel to the long edge of the block. FH is consequently to be at right angles to the block when one looks down upon the objects. In other words, FH is going the same way as the ends of the block, as also is GH, which is the trace on the ground-plane of a plane vertically downwards containing FH. GH and LK will vanish to the same point, somewhere on the left; and FH will vanish to an A.V.P. at that position over the V.P. of GH, which represents an inclination of 30°.

The problem works out as follows: First find V.P.'s of

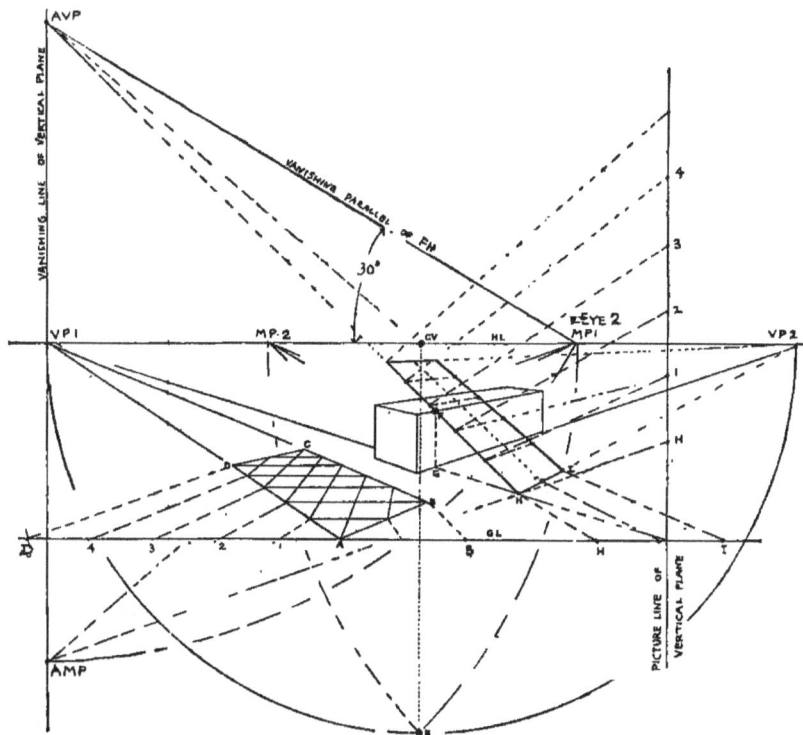

Fig. 184.—The problem solved.

the objects drawn in the diagram Q. 4. These V.P.'s are V.P.$_1$ and V.P.$_2$, and apparently they are at 45° to right and left, and occupy the positions of the distance-points. Find M.P.$_1$ and M.P.$_2$. With M.P.$_1$ measure AD and the four intermediate dimensions on the ground-line.

R

This gives the actual size of the side AD. With M.P.$_2$ measure side AB in the same way.

Proceed now to put the new rectangle upon the block. Drop a line from F to G, and draw a line through G from V.P.$_1$: upon this line the lower end of the rectangle will occur. As has been said above, the exact position of H is determined by the slant that FH is required to make. If we get the V.P. of the slanting line FH, we can draw from it through F and find H, where the line cuts the line previously drawn from V.P.$_1$ through G. To find the V.P. (or A.V.P.) of FH, we proceed as follows: FH will vanish over V.P.$_1$, because FH has to be over GH. The V.P.'s of FH and GH will therefore be on the V.L. of a vertical plane. We draw through V.P.$_1$ a vertical V.L.: somewhere up this line our A.V.P. of FH will occur. Now, the distance up is to be found by an angular measurement from the eye. The eye's position in relation to this vertical V.L. will be at M.P$_1$ (see page 109, where the method of obtaining the eye in relation to a vertical V.L. is given). The line (horizon) from eye$_2$ to V.P.$_1$ represents the level, and V.P.$_1$ will be the V.P. of all horizontal lines in the vertical plane. The vanishing parallel, representing 30° with the horizontal, will therefore be drawn from the eye$_2$ 30° above horizon. A.V.P. is therefore obtained, and a line drawn from it through F will find H, the lower limit of the one side of the rectangle.

Find A.M.P. by an arc struck from A.V.P. with compass stretched to eye$_2$. Next find the picture-line from the vertical plane containing FH. The vertical plane will cut through the ground. Now, GH is at the same time both in the vertical plane and in the ground-plane. We have the picture-line of the ground-plane ; indeed, it is the only one we have so far. We draw a line from V.P.$_1$ through GH, and bring it forward till it cuts the G.L. Where it does, we draw a vertical picture-line—the front edge of the plane containing FG and H, and receding to the V.L. on the left.

From A.M.P. we now draw a line through H to the new picture-line, and so get H upon it. Up from H we set off 1, 2, 3, etc., according as we have them on the ground-line in A1234D. From these points we take measuring-lines back to A.M.P., and so get the dimensions on the edge FH of our new rectangle.

HI is vanished to V.P.$_2$, and is measured on the G.L. by M.P.$_2$. The farther long edge of the rectangle is then vanished to A.V.P. The middle of HI is found by M.P.$_2$, and it also is carried up the rectangle to A.V.P. The lateral lines are then carried across from the front edge FH to V.P.$_2$, and the skeleton of the rectangle is complete.

The pattern is not drawn on the rectangle in the diagram, as clearness is retained, and there is no great knowledge or skill required in drawing it. The student should find the V.P.'s of these diagonal lines. They will vanish in a V.L. of the oblique plane of the rectangle.

QUESTION 5 (June, 1901).

*Diagram Q. 5 shows in perspective a rectangular oblique plane rising from the ground-plane to an upper horizontal plane. On the oblique plane is a point A. The horizon, centre of vision, distance of the eye from the picture, and ground-line are given. At point A erect a vertical line 10 feet high to represent the nearest edge of*

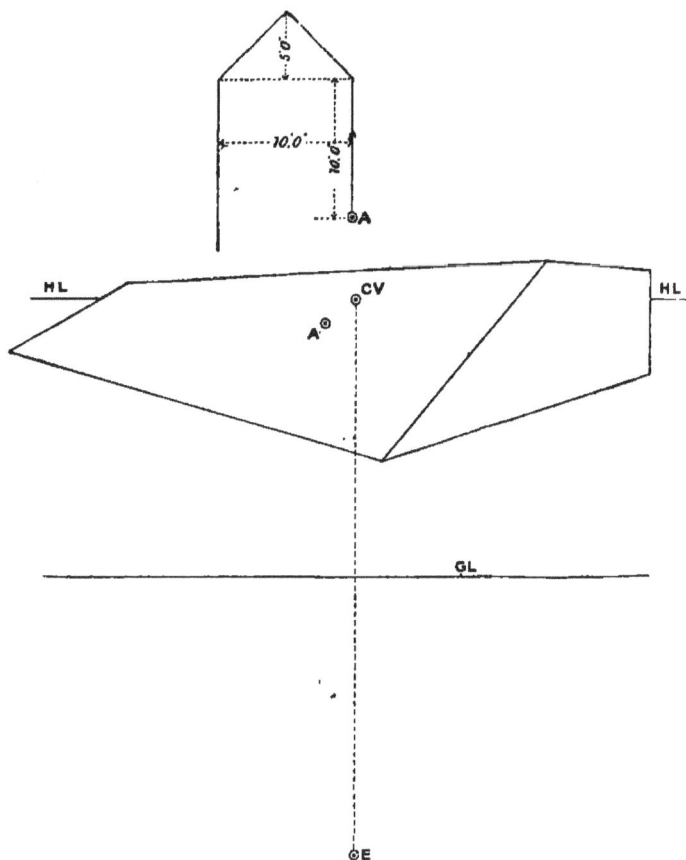

Fig 185 —Diagram Q. 5. (The scale is ¼ inch to 1 foot.)

*a house rectangular on plan and standing on the oblique and upper horizontal planes. The gable end of the house is to be of the dimensions shown in the small diagram, and to be in a vertical plane inclined 30° to the picture towards*

*the left. The side wall of the house is to be 20 feet long
Scale, ½ an inch to a foot.* (32 marks.)

*The horizon must be drawn across the short way of the
paper, 6 inches from the top, and the centre of vision
placed 5¾ inches from the right-hand edge of the paper.*

The subject of this exercise will evidently be something like the sketch, Fig. 186. That is to say, a house
vanishing at 30° to left
and 60° to right is
planted upon a bank at
a different angle, pro-
bably 45° to right and
left. We have to make
the house of certain di-
mensions, which are given,

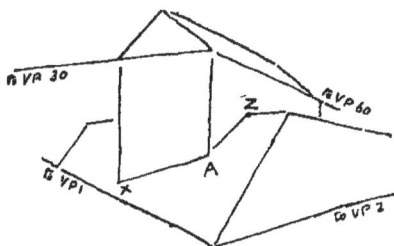

FIG. 186.—Probable aspect of the
problem.

and we have to begin by making the edge over A 10
feet in height. The construction of all the upper part
of the house should not give much trouble, but the
junction of house and bank will introduce a difficulty.

To begin with, we have to find the V.P.'s and V.L.
of the oblique plane. This is an easy matter, for we
merely continue the edge of the bank to the horizon,
and there find V.P.₁ and V.P.₂. Over V.P.₂ we raise
a vertical V.L. to contain the A.V.P. of the slope of
the bank. We continue the slanting edge BD upward,
and find A.V.P. The line from B to A.V.P. is over the
line from B to V.P.₂. We can now draw our V.L. of
oblique plane through V.P.₁ and A.V.P.

To raise 10 feet over point A, we proceed thus. We bring down a line from A.V.P. through A. This gives E; then through E we bring forward a line from V.P.₂ under A, right out to the G.L. at F. At F. we make a line of height FG 10 feet high, and take a line back over FE to V.P.₂; this gives H over E, so that EH is 10 feet. Then from H we draw a line to A.V.P. parallel to EA, and so get HI with I 10 feet over A.

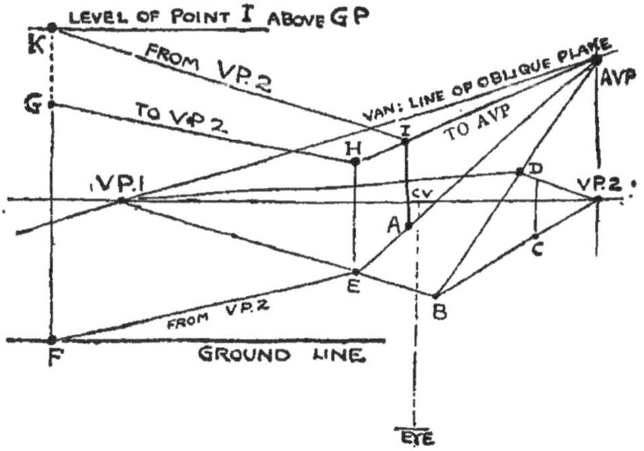

Fig. 187.—The V.P.'s and V.L. of the oblique plane of the bank. The edge 10 feet raised above A. The level of I found on the picture-plane.

If now we lengthen the line of height FG upward, and bring a line through V.P.₂ to cut it, we get K, which is the level of I in relation to the ground-plane. K is, of course, in the picture-plane, so a horizontal line through K will give us the picture-line or new ground-line of a horizontal plane through I. The eaves of the house will be in a horizontal plane, and this will be its P.L.

The eaves of the house are in a horizontal plane.
This horizontal plane has its V.L. at the horizon; its
P.L. is through K, as we have just seen; its eye is the
usual eye down below. The gable end has to go at
30° to left, and the side at 60° to right. We find these
V.P.'s (Fig. 188) by vanishing parallels, and also get
the M.P.'s in the usual way.

Point I is our starting place. We draw to V.P. 30°

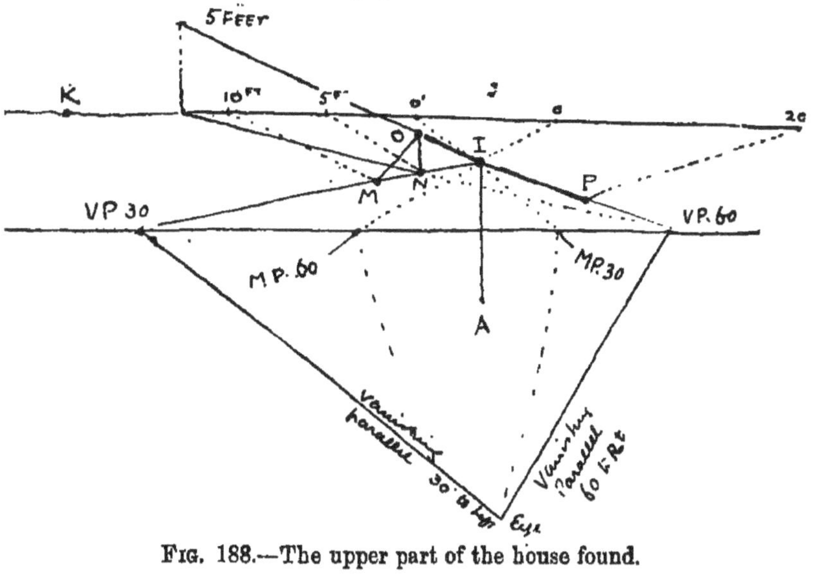

FIG. 188.—The upper part of the house found.

from I, and along it measure by M.P. 30°, the distances
across the gable—that is, 10 feet for the full width, and
5 feet for the middle. We thus get point M and N.
Over N we now want a vertical height of 5 feet. We
make a wall of height by bringing a line from V.P. 60
through N to the P.L., raise our 5 feet, and take a line
back to V.P. 60; where a line up from N cuts this line
is point O, the apex of the gable.

From I a line to V.P. 60 gives the side. It is measured from M.P. 60.

Note particularly that the line of height for getting the 5 feet up from N to O is not the same line of height as was previously used for FG and K. That line goes to V.P.$_2$; this goes to V.P. 60.

The junction between house and bank can be found in two ways.

The student will remember that in Fig. 187 the line from E to V.P.$_2$ is under A. Consequently, if a line

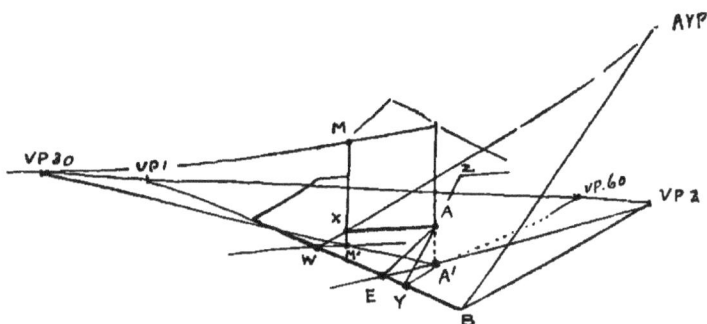

Fɪɢ. 189.—The junction obtained by means of the ground-plane.

be dropped from A it will find what we may call the foundation of the house on the ground-plane at A'. Now, the plane of the *side* of the house does not pass through V.P.$_2$, which is the V.P. concerned in EA', but we can make A'Y from V.P. 60, which is the V.P. concerned in the side of the house. YA will be where the plane of the house cuts the bank, and YA can thus be continued up to the top at Z.

We do not already possess point X, as we do A; we only have M above it. But we can find M' on the ground by drawing from A' a line under IM. IM vanishes at 30° left, so A'M' must vanish there too, *not to* V.P.$_1$. A line being thus drawn from A' to V.P. 30, and a vertical dropped from M, we get M'. Just as A' is related to E and E to A, so M' can be related to W and W to X by drawing through M' a line from V.P.$_2$ to the edge of the bank, and then another from W to A.V.P., giving X.

Another method is to find the V.P.'s of the lines of

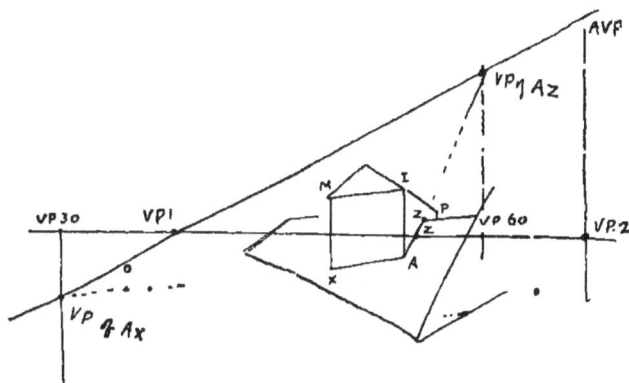

Fig. 190.—The junction found by the V.P.'s of the lines of junction.

junction. This is done in Fig. 190. We arrive at the method as follows: IM vanishes at V.P. 30. AX is under IM; it is in the same vertical plane, therefore the V.P. of AX will be on the same vertical V.L. as IM. Now, the V.P. of IM is V.P. 30. Through V.P. 30 we draw a vertical V.L. But AX is also in the oblique plane, and must have its V.P. on the V.L. of the oblique plane. Its V.P. will therefore be at one and the same time in both the vertical V.L. through V.P. 30 and the oblique V.L. Where the two V.L.'s cross must be the

V.P. of AX. Having point A, we can easily draw to the V.P. of AX, and so obtain X by a line falling from M.

In the same way, AZ is in both the vertical plane through V.P. 60 and the oblique plane. Its V.P. must therefore be where the vertical V.L. through V.P. 60 strikes the oblique V.L.

### QUESTION 6 (June, 1901).

*Diagram Q. 6 gives the perspective view of a block, rectangular on plan, and of a vertical reflecting surface RS, which is at an angle with the picture, both standing on the ground-plane. The horizon, centre of vision, and distance of the eye from the picture are given. Draw the reflection of the block on RS.* (30 marks.)

*The horizon must be drawn across the short way of the paper, 5 inches from the top, and the centre of vision placed 5 inches from the right-hand edge of the paper.*

The image seen in a mirror is the counterpart of the object of which it is the reflection. The image is related to the mirror in precisely the same way as the object. The situation is always best seen if a sketch be made of the object and the mirror when the mirror is reduced to a line. In the case before us these conditions point to a plan, and in plan we see that the object has AB and CD level or parallel to the picture, AC and BC perpendicular to it. These facts are deduced from the perspective drawing supplied in diagram Q. 6. The image is A'B'C'D', and A'C' strikes the reflecting surface at the same angle as AC; CC′ is thus perpendicular to the reflecting surface, and so are AA′ and DD′.

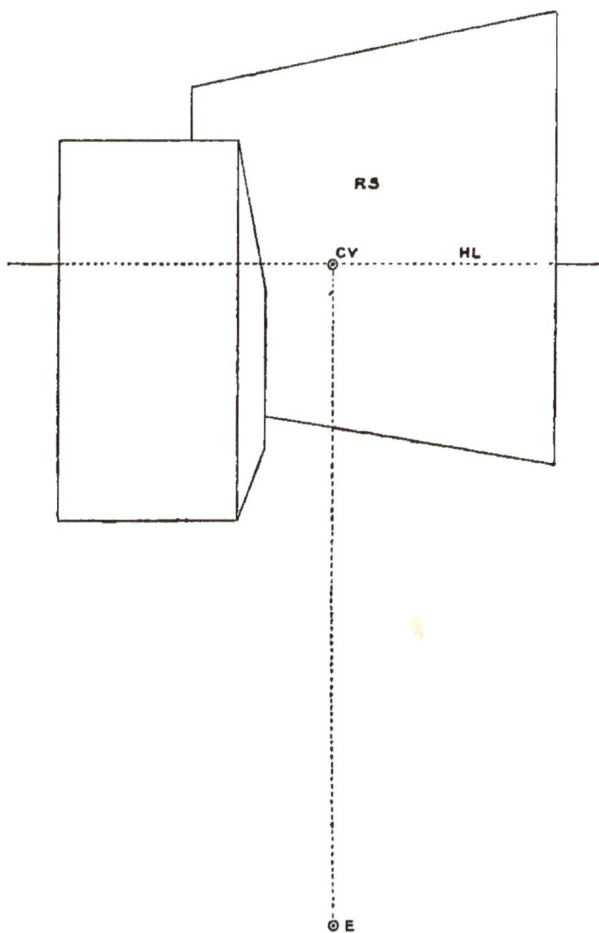

FIG. 191.—Diagram Q. 6.

We have, in working the problem, merely to put into perspective the image as if it were a second object, vanishing it to such V.P.'s as the laws of reflection determine. These laws also tell us that C′ is as far within the mirror as C is before it, and so on of all the points.

The angle of incidence is equal to the angle of reflection, therefore if BA makes a certain angle of

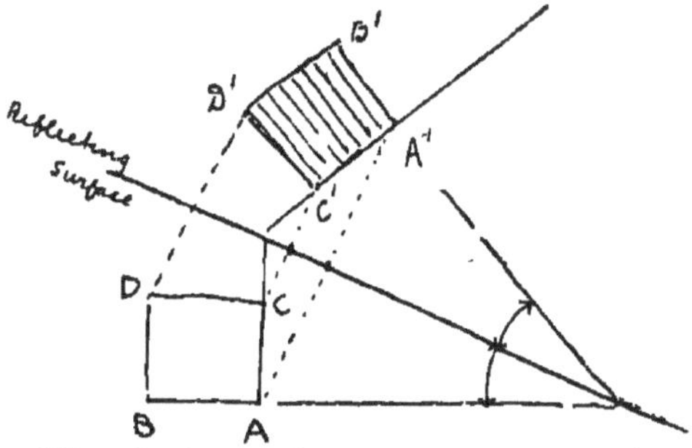

Fig. 192.—A plan of the object, mirror, and image.

incidence on this side of the mirror, its image will make the same angle as an angle of reflection on the other side. Hence, if (Fig. 193) 3 is the vanishing parallel of the mirror, and 1 the vanishing parallel of AB, then 4 will be the vanishing parallel A′B′, because the angle between 4 and

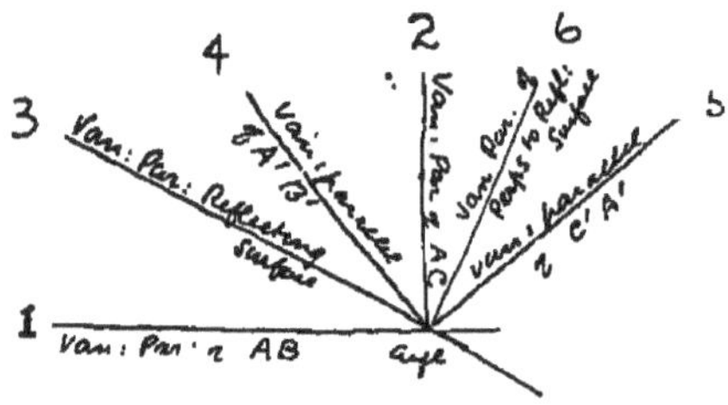

Fig. 193.—The various vanishing parallels.

3 is the same as between 3 and 1. AC vanishes to C.V.: where will its reflection vanish? The vanishing parallel of the reflection of AC will be 5. A vanishing parallel must always proceed away from the eye forward. Putting down the angle equal to the angle between 2 and 3, it comes below 1, and so must be continued up through the eye into 5. This is shown in the solution

Fig. 195. We must also have a vanishing parallel for the lines perpendicular to the mirror. These are C, C', etc., and their vanishing parallel is 6.

The top of the object is not level, but slants down, its slanting edges vanishing below C.V. in A.V.P., shown in Fig. 194. The vertical line through C.V. is the V.L.

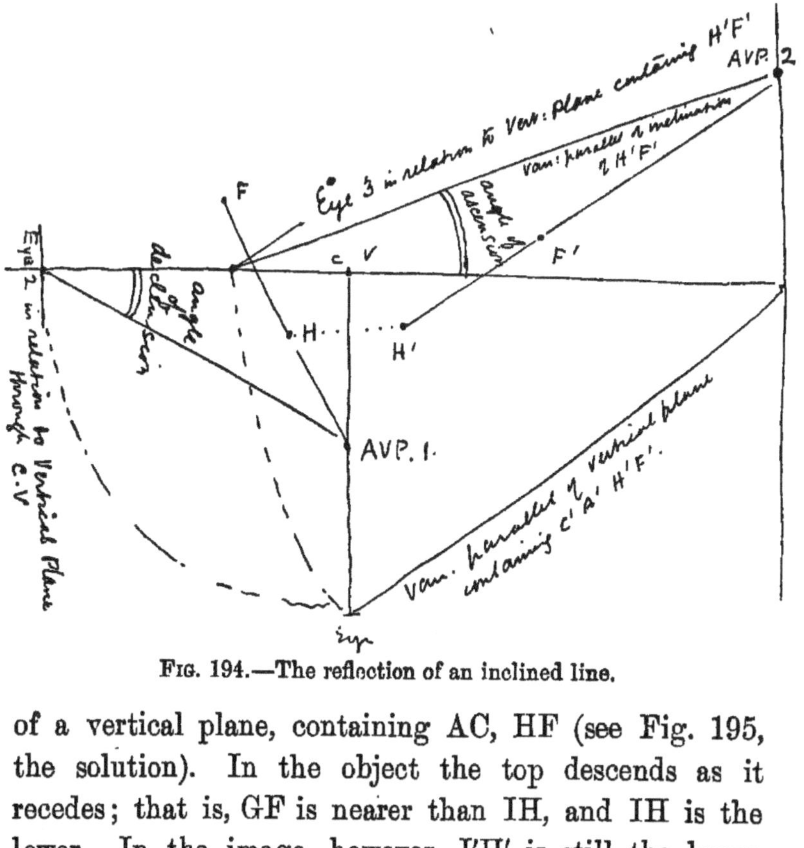

FIG. 194.—The reflection of an inclined line.

of a vertical plane, containing AC, HF (see Fig. 195, the solution). In the object the top descends as it recedes; that is, GF is nearer than IH, and IH is the lower. In the image, however, I'H' is still the lower, but it is the nearer, and so the top *ascends* as it recedes. If the V.P. of FH is below the V.P. of AC, the V.P. of H'F" will be above the V.P. of C'A'. The V.P. of C'A' will be far out on the right. Our paper does not permit

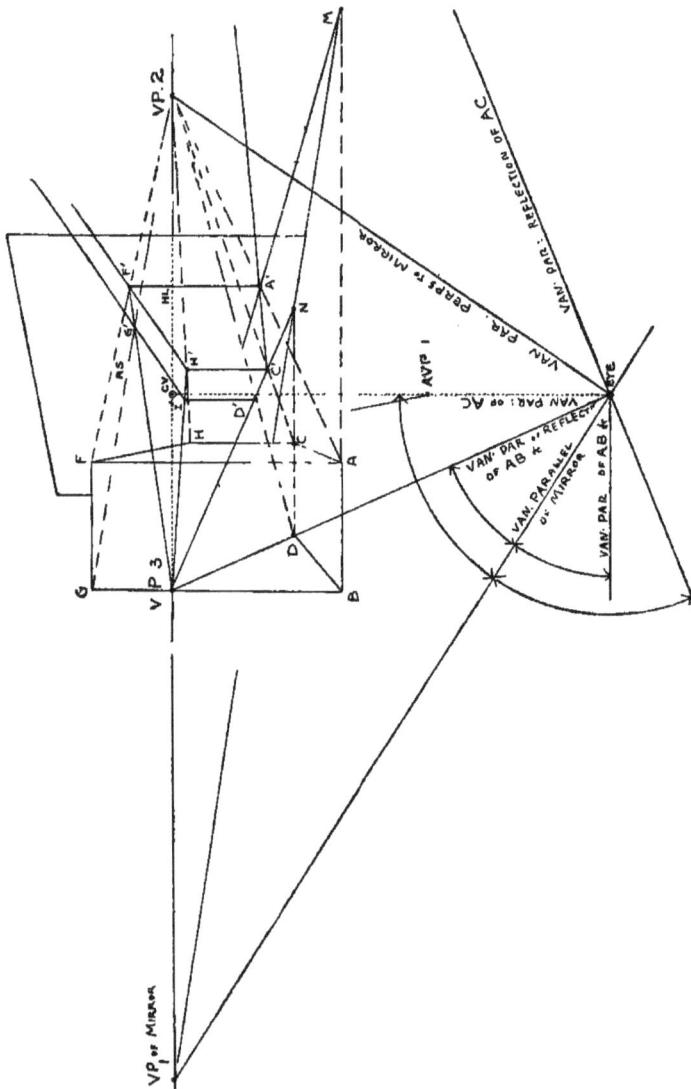

FIG. 195.—The solution of the problem.

our getting these points, but if it did we should proceed as in Fig. 194. Here FH is the declining line vanishing

to A.V.P.$_1$. We have to find at what angle it is descending. Eye 2 (position of DP) is the eye for the plane
through C.V., and we readily find the angle of declension.
Our reflection H'F' is in a plane far out on the right.
We draw the vanishing parallel from eye 1, get the
vertical V.L., then eye 3, and put up the same angle
from it to find A.V.P.$_2$, as we found was that at which
A.V.P.$_1$ was declined below the horizon. A.V.P.$_2$ is the
V.P. of the reflection of FH, supposing the mirror to be
where the question here treated of places it.

Not being able thus to get the V.P. of the reflection
of AC, we work the problem by AB and CD. Continue
both these to the mirror, M and N, prolonging the edge
of the mirror for the purpose. The reflection of M'A will
vanish to V.P.$_3$, and A' will be located by a line from A
to V.P.$_2$. The lines from the various points of the object
to V.P.$_2$ are freely used, the upper part of the reflection
being found entirely by them, and verticals from A'C'
and D' below. G' is, however, found by vanishing F'G'
to V.P.$_3$.

### QUESTION 7 (June, 1901).

*Diagram Q. 7 shows in per-*
*spective a cube standing on a*
*horizontal plane. Find the*
*vanishing points, horizon,*
*measuring points, centre of*
*vision, and distance of the eye*
*from the picture.*

(5 marks.)

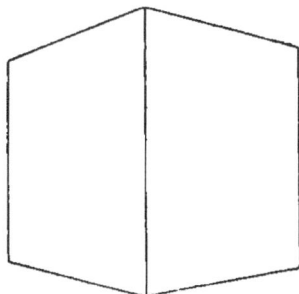

FIG. 196.—Diagram Q. 7.

This question is practically the same as No. 1 of April, 1901. Being in this case a cube, no sizes of sides need be stated, for all the sides of a cube are equal, and the

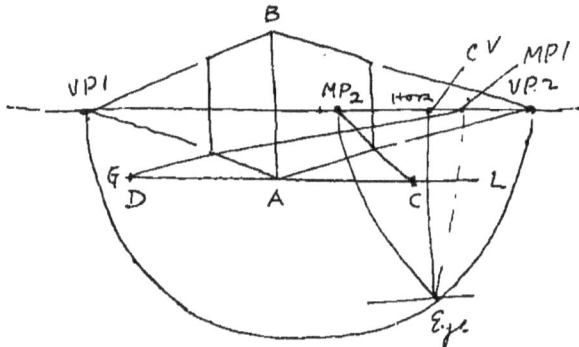

FIG. 197.—A sketch solution of the problem.

nearest vertical AB sets the standard. The G.L. is drawn through A, so that AB is assumed to be in the picture and actual. AC, therefore, is the length of AB; so is AD.

QUESTION 8 (June, 1901).

*Diagram Q. 8 shows in perspective a square slab lying on a horizontal plane. The horizon, centre of vision, and distance of the eye from the picture are given. Show how the aspect of the slab would be changed if it were revolved on a pin passed through the holes A and B (the centre points of its lower and upper surfaces), so that its edges are inclined at angles of 45° with the picture-plane.*

The slab is in parallel perspective; two edges are parallel to the picture and two vanish to C.V. The new slab has to have its edges at 45°. Now, the V.P.'s at 45° are not only the distance points for the V.C., but they

are also the V.P.'s of the diagonals of a square. They
are, therefore, easily found. In the new slab the diagonal

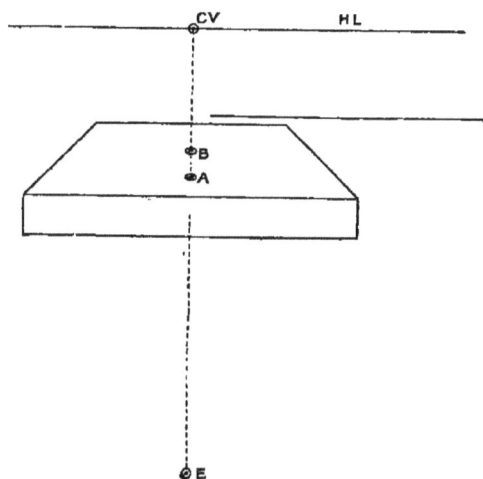

Fig. 198.—Diagram Q. 8.

through B will be parallel to the horizon, and will be seen
at its full length. If we assume the portion of the front
edge of the first slab GCFD as in the picture, GC can act

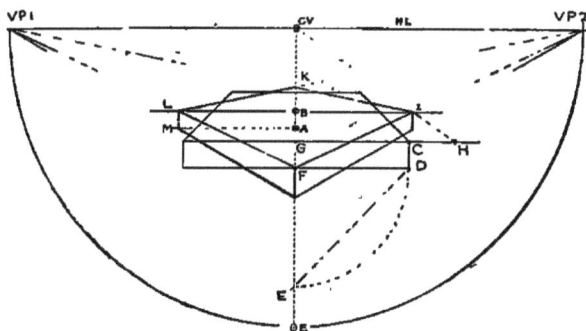

Fig. 199.—The solution of the problem.

as a ground-line. With FD as a side we can construct a
small square ($\frac{1}{4}$ of the slab, and so get DE) the actual

S

length of half a diagonal of the slab. From G we set off GH equal to DE, and draw from H to CV. This yields

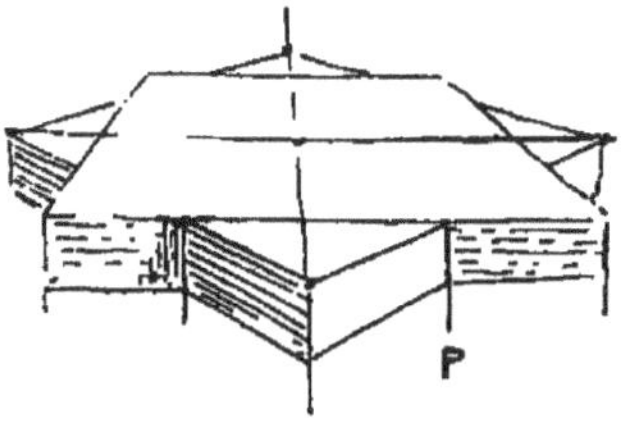

FIG. 200.—A mistake in the solution corrected.

I on the diagonal through B. By means of V.P.$_1$ and V.P.$_2$ we readily complete the figure.

Through some slight error or errors in the working such as constantly occur, the two slabs are not fitting together as they should. This error is corrected in Fig. 200, where the junction P is vertical, as it should be. This slight matter shows how valuable the sketched perspective is, for here is a case where slight errors of execution result in a mistake which would never occur in the freehand sketch.

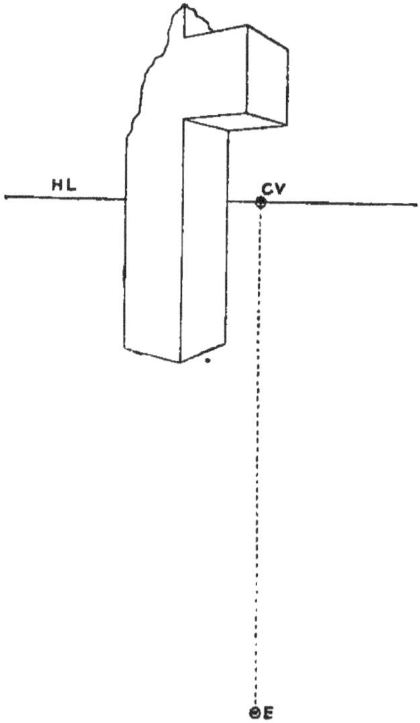

QUESTION 9 (June, 1901).

*Diagram Q. 9 is the per-spective representation of a mutilated cross. The hori-zon, centre of vision, and distance of the eye from the picture are given. Restore the missing arms of the cross to the length of the complete one.* (9 marks.)

FIG. 201.—Diagram Q. 9.

In the solution of this question we first find V.P. of

AB, the front edge of base of cross. Then we bring BA forward by means of this V.P.₁ until we can cut it off in C' under C. Through C' we make our G.L. Finding M.P.₁, we find A' and B' on the ground-line, A'C' being the actual length of an arm of the cross. We therefore

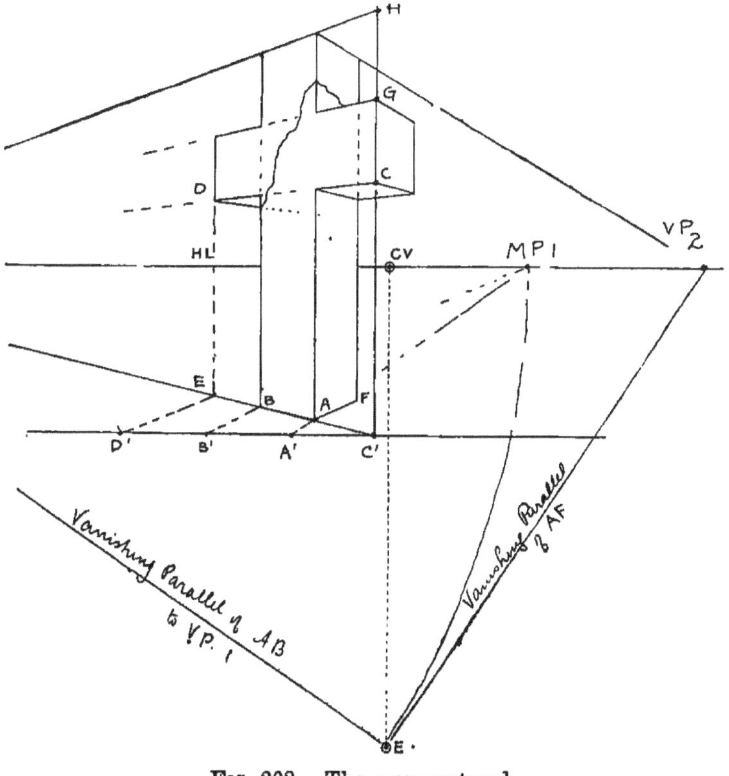

Fig. 202.—The cross restored.

set off B'D' equal to A'C', and obtain E, which, when thrown up, meets the lines from G and C, and completes the left arm of the cross. The under side of this new arm vanishes to V.P.₂. The top is made by continuing the line of height C'G to H, making GH equal to A'C'.

## QUESTION 10 (June, 1901).

*Diagram Q. 10 gives a perspective view of a horizontal road, between two planes inclined to it, the one on the right upwards, that on the left downwards till it meets a lower horizontal plane, on which, at point B, is a tree.*

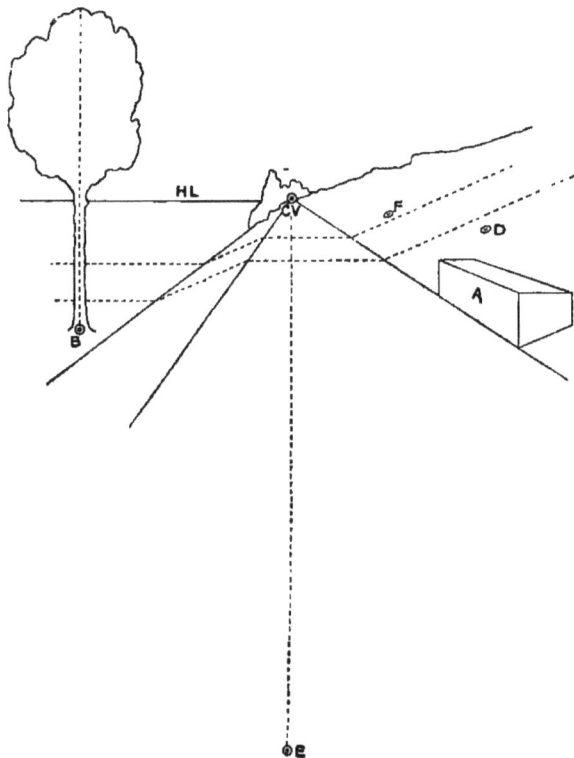

FIG 203.—Diagram Q. 10.

*The bench A is 8 feet long. The horizon, centre of vision and distance of the eye from the picture are given. Find the height of the tree, stating the dimension in figures, and at points D and F on the inclined plane erect vertical lines to represent trees of the height of that at B.*

(11 marks.)

The bench is 8 feet long. To find a scale we shall have to get DP, along the horizon from C.V., the distance C.V. to eye. Bringing a line down from DP through the far end of the bench, we get the size shown upon the

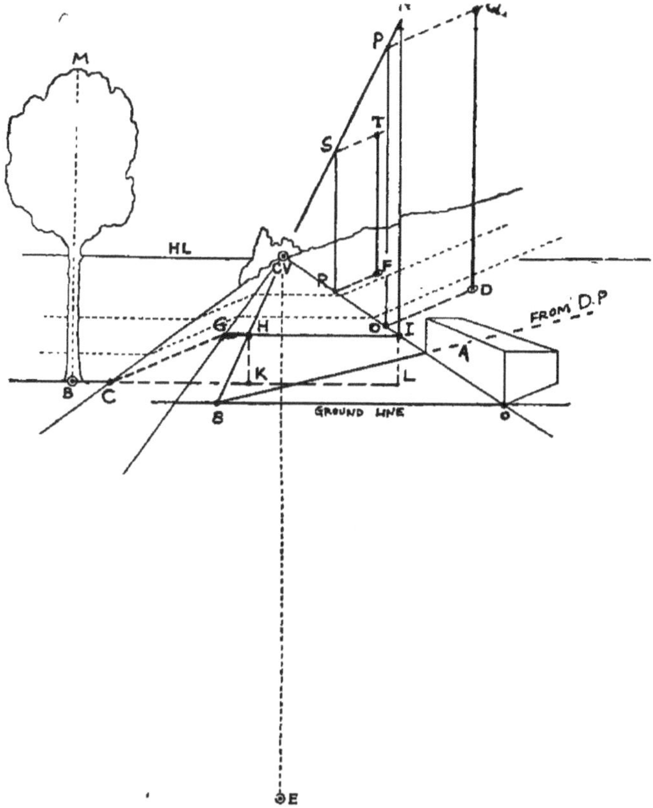

FIG. 204.—The problem solved.

ground-line, which we draw through the near corner of the bench. 8,0, is then 8 feet, and if we like, we can divide it into 8 parts and make a scale.

We have now to get the tree in relation to this scale.

The road and banks recede to C.V., and therefore the

perpendiculars to them are parallel to the picture, and are drawn either with T square or set square. We draw with the T square from B to C, then parallel (for the slanting lines are found not to vanish) with the slanting lines CG, and so obtain a point on the edge of the road level with the tree. Across the road we draw GI, and we cross this by a line from 8 on the ground-line to C.V. This gives H. HI is therefore 8 feet. HI is in the same parallel as the tree, aud we can measure our tree from HI. But if we like we can drop the size HI down to KL, level with the foot of the tree. By applying the dividers it will be found that BM is just double KL, so that BM is 16 feet.

To get our two other trees, we proceed thus. HI is 8 feet. IN is a vertical 16 feet, just double HI. From N we draw a line to C.V. This marks the height of 16 feet all along the edge of the road. Next we bring slanting lines, parallel to the other slanting lines, down from D to O and F to R; at O and R raise verticals to P and S. OP and RS are both 16 feet. Then by more slanting lines we carry P over D to Q, and S over F to T. DQ and FT are the required trees.

QUESTION 11 (June, 1901).

*Diagram Q. 11 shows in perspective a wayside inn, a man at point A on the ground-plane, and a staff with flag flying. All the details of the house are accurately drawn, its nearest edge BD being 12 feet long. The horizon, centre of vision, and distance of the eye from the picture are given. Find the errors in the perspective and indicate the methods for rectifying them. The man is*

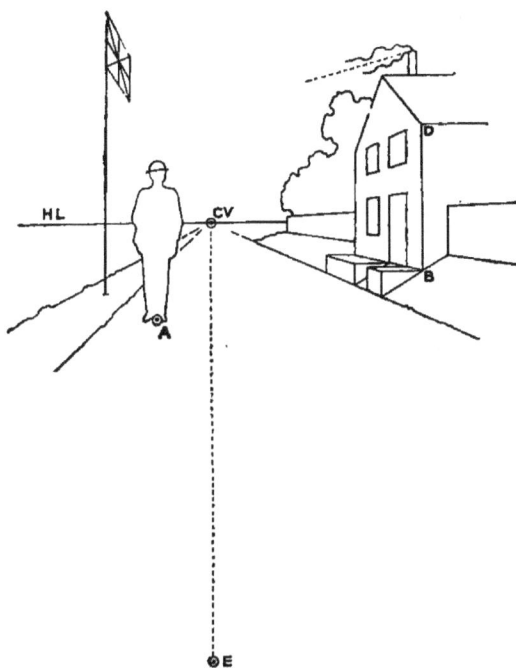

Fɪɢ. 205.—Diagram Q 11.

*supposed to be 6 feet high, and the flag and smoke to be affected equally by the wind.* (12 marks.)

If BD is 12 feet high, then, as it is a vertical line and does not vanish, half of it must be 6 feet, the height the man should be. We therefore find the halfway up BD. The house vanishes to C.V., so its sides are parallel to the picture, and the slanting bank will have its lines not vanishing, but drawn with the set square. FB is such a line. We therefore draw from the point, which is

6 feet above B, a slanting line parallel to FB. This
gives us E and EF, 6 feet high, for in parallel perspective
B is no further away than F. We now carry this height
of 6 feet forward over the edge of the road by a line
from C.V. We transfer point A to the edge of the
road G, and, finding by a vertical GH, 6 feet, we

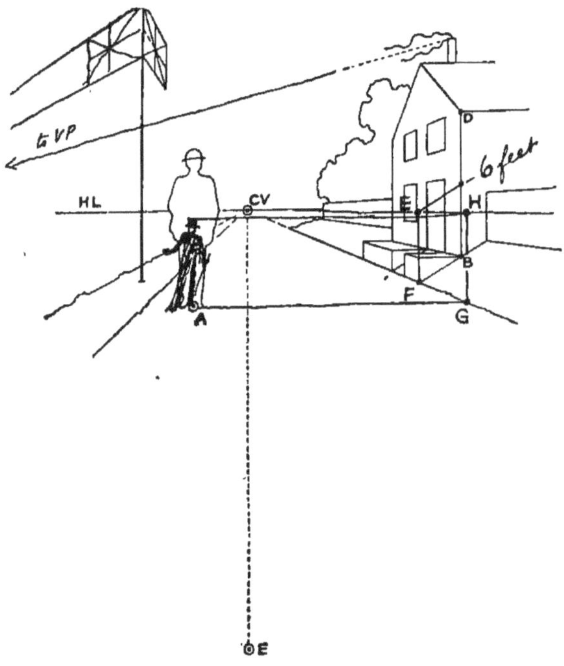

Fig. 206.—The problem solved.

transfer H back over A, which gives the correct height
of the man.

Assuming that the smoke is following a level course,
it is vanishing on the H.L. somewhere on the left. By
continuing the dotted lines in the smoke this V.P. is
found, and the flag made to vanish there also.

QUESTION 12 (June, 1901).

*Diagram Q. 12 shows in perspective a wall and door-way, and at points A and B on the ground-plane vertical lines, that at A throwing a shadow AD. The horizon.*

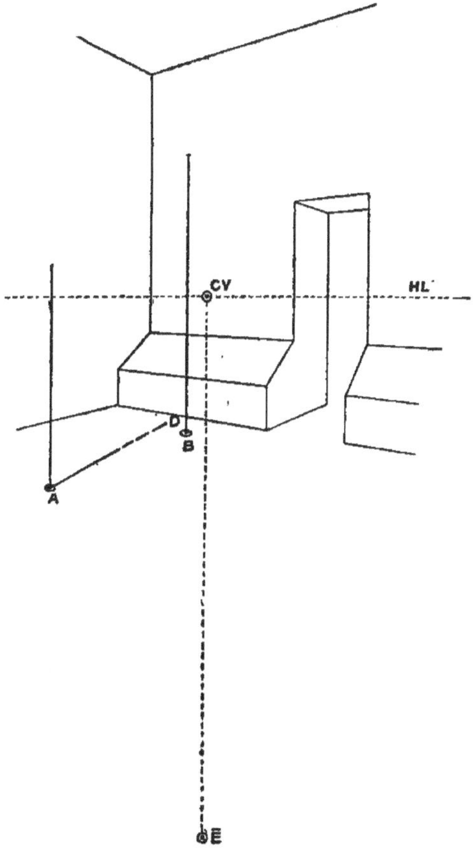

FIG. 207.—Diagram Q. 12.

*centre of vision, and distance of the eye from the picture are given. Find the vanishing point of the sun's rays and draw the shadows of line B and of the doorway.*

(15 marks.)

*This drawing will occupy a space* 15 *inches wide and* 10½ *inches high, in the centre of which place* O.V.

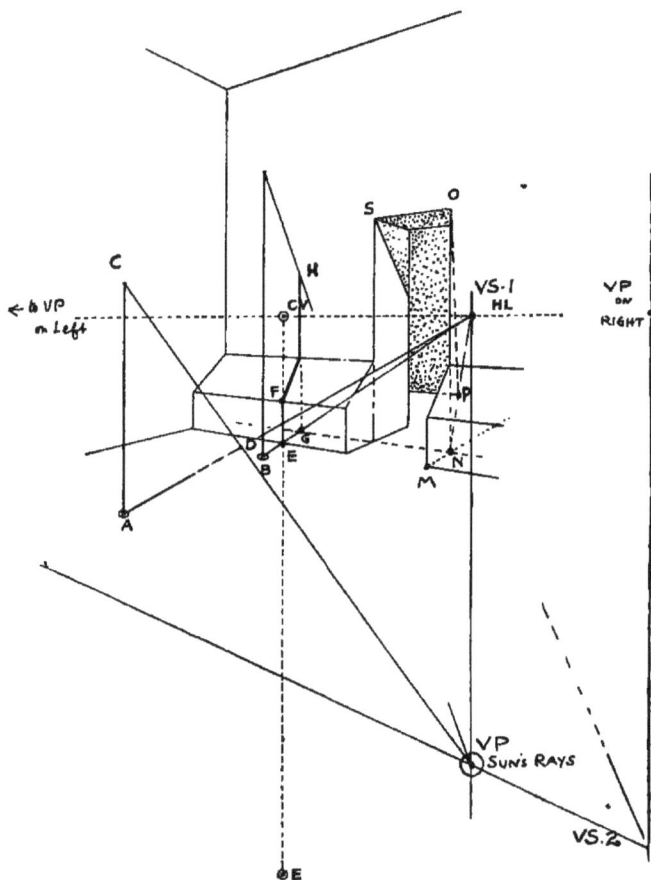

FIG. 208.—The solution of the question.

In this problem the shadow of the jamb of the doorway is hidden, and the shadow of only the top line of the doorway is seen.

# RULES

FOR PARALLEL PERSPECTIVE.

Horizon (Hor.).   Drawn with ⊤ square anywhere.

C.V.   Marked anywhere on horizon.

Eye.   Distance of spectator before picture plane (length of P.V.R.), placed from C.V. up or down at right angles to horizon.

D.L. (Directing-line).   Drawn through eye parallel to horizon.

G.L. (Ground-line).   Front edge of ground.   Drawn with ⊤ square parallel to horizon, as far in actual feet below it as it is said to be, or as spectator is high.

P.L. (Picture-line).   Front edge of any plane not ground.   Found in same way as G.L.

C.V. is V.P. (vanishing-point) of lines directly receding.

D.P. (Distance-point) or M.P. of C.V.).   With C.V. as centre, and distance C.V. to eye as radius, describe arc to cut horizon.

*To use the above—*

Along G.L. or P.L. measurements are actual, and the sizes are taken from the scale.   Lines from these dimensions back to C.V. convey *width* back into the picture.

*Height* is found by erecting lines upon the G.L. or P.L. and carrying the dimensions (marked on them from the scale) by lines backward to C.V.

*Distance* within the picture.   Take any line from G.L. to C.V. along G.L., from commencement of line upon it, set the distance required, and from the end of it draw to D.P. crossing the line to C.V.   Lines to D.P. from points on G.L. or P.L. transfer the distances between those points on to lines vanishing at C.V.

FOR HORIZONTAL PERSPECTIVE.

*The same as for parallel perspective, with following additional :—*

V.P. (vanishing-point) of lines not vanishing to C.V. but somewhere else on horizon.   Set up on D.L. at eye the angle at which the line recedes and continue till horizon is reached.

N.B.—The two V.P.'s of a rectangle will at the eye have 90° between their " vanishing parallels."

M.P. (measuring-point) of a V.P. With centre at V.P., and radius equal distance of V.P. from eye strike an arc to cut horizon. Dimensions placed on G.L. or P.L. and carried to M.P. by lines will be transferred to any line going to the V.P. from which the M.P. was made.

FOR INCLINED PERSPECTIVE.

*The same as above with the following—*

V.L. (Vanishing-line—similar to horizon) of five kinds of planes in addition to the horizontal can be used.

*Of planes perpendicular to picture—V.L. through C.V.*

(i.) Vertical. Drawn with T square through C.V.

(ii.) Inclined. Draw through C.V. a line making with horizon the angle of inclination.

*Of planes inclined to the picture—V.L. not through C.V.—*

(iii.) Vertical. Find V.P. (as in horizontal perspective) of direction, and draw with T square vertically through it.

(iv.) Directly ascending or descending. Find V.L. of such a plane as (i.) above at its eye (see below) set up on horizon the angle of inclination to cut the V.L. over C.V. Draw horizontally with T square.

(v.) Obliquely ascending or descending. Find on horizon V.P. of intersection with horizontal plane. Find on horizon V.P. of direction of inclination at right angles to V.P. of intersection. Through V.P. of direction draw vertical V.L. (as iii. above). Find eye for this V.L. (see below) At eye erect angle of inclination to cut vertical V.L.. and thus finding V.P. of inclination. Through this V.P. and the V.P. of intersection draw the V.L. required.

EYE FOR NEW V.L.'s.

*Planes perpendicular to picture—V.L. through C.V.—*

(i). Vertical. The D.P. is the eye.

(ii.) Inclined. The eye is as far from C.V. on a line perpendicular to V.L. as D.P. is from C.V.

*Planes inclined to the picture— V.L. not through C.V.—*

(iii.) Vertical. The M.P. for the V.P. of direction found as in horizontal perspective is the eye.

(iv). Directly ascending or descending (see page 112).

(v.) Obliquely ascending or descending (see page 112).

Picture-lines of the various planes. Found by first crossing the V.L. of a plane whose P.L. one has (as horizon and G.L.) by

V.L. of plane whose P.L. one requires. If the two V.L.'s cannot cross a third, crossing both, must be assumed as a link between them.

A.V.P. and A.M.P. Found from the new eye precisely as in horizontal perspective.

ADOWS.

*Artificial Light—*

R.S. (Radiating point of shadow). Draw a line from the light parallel *in fact* to the line whose shadow is required, and where the line cuts the plane on which shadow falls is R.S.

   N.B.—The lines drawn parallel *in fact* have to vanish to the V.P. of the line if such exist.

Direction of shadow. Drawn by line from R.S.

Length of shadow. Determined by a ray drawn from light through end of line casting shadow.

*Shadows cast by the Sun—*

V.P. of sun's rays. Is a V.P. in a vertical plane (as iii. above) and is found by using the new eye and the V.L. of the vertical plane.

   N.B.—When sun is *behind* spectator (same as *before* picture) the rays vanish downward, when sun is *before* spectator they vanish upward.

V.S. (Vanishing-point of shadow).

V.S. of line *parallel* to picture with sun's rays *parallel* to picture. No V.S. but shadow drawn parallel to V.L. of plane receiving shade.

V.S. of line *parallel* to picture with sun's rays *at angle* to picture. Draw through V.P. of sun a line parallel to line casting shadow. Where this line cuts V.L.'s of planes receiving shade are V.S.'s.

V.S. of line *inclined* to picture with sun's rays *parallel* to picture. Draw through V.P.'s of lines lines parallel to direction of sun's rays. The lines to cut the V.L.'s and so yield V.S.'s.

V.S. lines *inclined* to picture when sun's rays are *inclined* to picture. Draw a line through V.P. of sun and V.P. of line. Where this line cuts V.L.'s of planes receiving shadow are V.S.'s.

   N.B.—If the plane receiving shadow be *parallel* to the picture, and has no V.L. on which V.S.'s can occur—the shadows are drawn parallel to a line through V.P. of sun and V.P. of line casting shadow.

   N.B.—Problems which give special difficulty may always be solved or proved by considering the extremities of lines as summits of vertical lines and getting the shadows of the vertical lines.

REFLECTIONS.

A simple general rule. Draw from the points or corners of the object, lines *perpendicular* to the plane of the mirror, carrying the lines *to* the mirror and noting where they touch it, and carrying them on, through, and beyond it, and then measuring, by perspective methods on these lines, points as far beyond the mirror as the points of the object on the same lines were before it. Perpendiculars to the various planes are treated in page 129, etc.

# INDEX

THE END

www.ingramcontent.com/pod-product-compliance
Lightning Source LLC
Chambersburg PA
CBHW021945220326
41599CB00012BA/1181